THE

ZOOLOGY

OF

THE VOYAGE OF H.M.S. BEAGLE,

UNDER THE COMMAND OF CAPTAIN FITZROY,

DURING THE YEARS

1832 TO 1836.

PUBLISHED WITH THE APPROVAL OF
THE LORDS COMMISSIONERS OF HER MAJESTY'S TREASURY.

Edited and Superintended by

CHARLES DARWIN, ESQ. M.A. F.G.S.

CORRESPONDING MEMBER OF THE ZOOLOGICAL SOCIETY,

AND NATURALIST TO THE EXPEDITION.

MAMMALIA,

BY GEORGE R. WATERHOUSE, ESQ.

CURATOR OF THE ZOOLOGICAL SOCIETY OF LONDON, ETC. ETC.

British Library Cataloguing-in-Publication Data
A catalogue record for this book is available from
the British Library

THE

ZOOLOGY

OF

THE VOYAGE OF H.M.S. BEAGLE,

UNDER THE COMMAND OF CAPTAIN FITZROY, R.N.,

DURING THE YEARS

1832 TO 1836.

PUBLISHED WITH THE APPROVAL OF
THE LORDS COMMISSIONERS OF HER MAJESTY'S TREASURY.

Edited and Superintended by

CHARLES DARWIN, ESQ. M.A. F.R.S. Sec. G.S.
BY
NATURALIST TO THE EXPEDITION.

PART II.
MAMMALIA,
BY
GEORGE R. WATERHOUSE, ESQ.
CURATOR OF THE ZOOLOGICAL SOCIETY OF LONDON, ETC. ETC.

LIST OF PLATES.

LIST OF PLATES.

LIST OF PLATES.

[*] The palatine foramina are accidentally omitted—see description.

GEOGRAPHICAL INTRODUCTION.

THE object of the present Introduction, is briefly to describe the principal localities, from which the Zoological specimens, collected during the voyage of the Beagle, were obtained. At the conclusion of this work, after each species has been separately examined and described, it will be more advantageous to incorporate any general remarks. The Beagle was employed for nearly five years out of England; of this time a very large proportion was spent in surveying the coasts of the Southern part of South America, and of the remainder, much was consumed in making long passages during her circumnavigation of the globe. Hence nearly the entire collection, especially of the animals belonging to the higher orders, was procured from this continent; to which, however, must be added the Galapagos Archipelago, a group of islands in the Pacific, but not far distant from the American coast. The localities may be briefly described under the following heads.

BRAZIL. This country presents an enormous area, supporting the most luxuriant productions of the intertropical regions. It is composed of primary formations, and may be considered as being hilly rather than mountainous. LA PLATA includes the several provinces bordering that great river;—namely, Buenos Ayres, Banda Oriental, Santa Fé, Entre Rios, &c. My collections were chiefly made at BUENOS AYRES, at MONTE VIDEO, the capital of Banda Oriental, and at MALDONADO, a town in the same province, situated on the northern shore, near the mouth of the estuary of the Plata. These countries consist either of an undulating surface, clothed with turf, or of perfectly level plains with enormous beds of thistles. Except on the banks of the rivers, trees nowhere grow ; there are, however, thickets in some of the valleys, in the more hilly parts of Banda Oriental. During the winter and spring of this hemisphere, a considerable quantity of rain falls, and the plains of turf are then everywhere verdant ; but in summer the country assumes a brown and parched appearance.

BAHIA BLANCA forms a large bay, in latitude 39° S. on a part of the coast, which falls within the territory of the province of Buenos Ayres, but which from its physical conditions would more properly be classed with Patagonia. The tertiary plains of PATAGONIA, extend from the Strait of Magellan to the Rio Negro, which is commonly assumed as their Northern boundary. This space of more than seven hundred miles in length, and in breadth reaching from the Cordillera to the Atlantic Ocean, is everywhere characterised by the dreary uniformity of its landscape. Nearly desert plains, composed of a thick bed of shingle, and often strewed over with sea-shells, (plainly indicating that the land has been covered within a recent period by the sea,) are but rarely interrupted by hills of porphyry, and other crystalline rocks. The plains support scattered tufts of wiry grass, and stunted bushes ; whilst in the broad flat-bottomed valleys, dwarf thorn-bearing trees, barely ornamented with the scantiest foliage, sometimes unite into thickets ; and here the few feathered inhabitants of these sterile regions resort. There is an extreme scarcity of water; and where it is found, especially if in lakes, it is generally as salt as brine. The sky in summer is cloudless, and the heat in consequence, considerable; whereas the frosts of winter are, sometimes, severe. The principal localities visited by the Beagle, were the RIO NEGRO, in latitude 41° S., PORT DESIRE, PORT ST. JULIAN, and SANTA CRUZ. At the latter place, a party, under the command of Captain FitzRoy, followed up the river in boats, to within a few miles of the Cordillera; and an opportunity was thus afforded of verifying the nature of the country in its entire breadth. At the Rio Negro the plains are much more thickly covered with bushes, (chiefly acacias,) than in any other part of Patagonia.

TIERRA DEL FUEGO may be supposed to include all the broken land south of a line joining the opposite mouths of the Strait of Magellan. The land is mountainous, and may be aptly compared to a lofty chain, partly submerged in the sea ;—bays and channels occupying the position of valleys. The Eastern side almost exclusively consists of clay-slate ; the Western, of primary, and various plutonic formations. The mountains, from the water's edge, to within a short distance of the lower limit of perpetual snow, are everywhere (excepting on the exposed western shores) concealed by an impervious forest, the trees of which do not periodically shed their leaves. On the East coast, the outline of the land shows that tertiary formations, like those of Patagonia, extend south of the Strait of Magellan ; but with the exception of this part, it is rare to find even a small space of level ground ; and where such occurs, a thick bed of peat invariably covers the surface. The climate is of that kind which has been denominated insular : the winters are far from being excessively cold, whilst the summers are gloomy, boisterous, and seldom cheered by the rays of the sun. In all seasons, a large quantity of rain falls. Hence, from the physical conditions of Tierra del Fuego, all the land animals must live either on the sea beach, (and in this class the Aborigines may be included) or within the humid and entangled forests.

The FALKLAND ISLANDS are situated in the same latitude as the Eastern entrance of the Strait of Magellan, and about 270 miles East of it. The climate is nearly the same as in Tierra del Fuego, but the surface of the land, instead of being as there, concealed by one great forest, does not support a single tree. We see on every side a withered and coarse herbage, with a few low bushes, which spring from the peaty soil of an undulating moorland. Scattered hills, and a central range of quartz rock, protrude through formations of clay-slate and sand-stone (belonging to the Silurian epoch,) which compose the lower country.

The structure of the west coast of South America, from the Strait of Magellan northward to latitude 38°, in its greater part, (as far north as Chiloe) is very similar to that of Tierra del Fuego. The climate likewise is similar,—being gloomy, boisterous, and extremely humid ; and, consequently, the land is concealed by an almost impenetrable forest. In the northern part of this region, the temperature of course is considerably higher than near the Strait of Magellan ; but nevertheless it is much less so, than might have been anticipated from so great a change in latitude. Hence, although the vegetation of this northern district presents a marked difference when compared with that of the southern; yet the zoology in many respects has, like the general aspect of the landscape, a very uniform character. The specimens were chiefly collected from the PENIN-SULA OF TRES MONTES, the CHONOS ARCHIPELAGO (from latitude 46° to 43° 30'), CHILOE with the adjoining islets, and VALDIVIA. The contrast between the physical conditions and productions of the East and West coasts of this part of South America is very remarkable. On one side of the Cordillera, great heavy clouds are driven along by the western gales in unbroken sheets, and the indented land is clothed with thick forests; whilst on the other side of this great range, a bright sky, with a clear and dry atmosphere, extends over wide and desolate plains.

CHILE in the neighbourhood of CONCEPCION (latitude 36° 42' S.) may be called a fertile land; for it is diversified with fine woods, pasturage, and cultivated fields. But towards the more central districts (near VALPARAISO and SANTIAGO) although by the aid of irrigation, the soil in the valleys yields a most abundant return, yet the appearance of the hills, thinly scattered with various kinds of bushes and cylindrical Opuntias, bespeaks an arid climate. In winter, rain is copious, but during a long summer of from six to eight months, a shower never moistens the parched soil. The country has a very alpine character, and is traversed by several chains of mountains extending parallel to the Andes. These ranges include between them level basins, which appear once to have formed the beds of ancient channels and bays, such as those now intersecting the land further to the south. North of the neighbourhood of Valparaiso, the climate rapidly becomes more and more arid, and the land in proportion desert. Beyond the valley of COQUIMBO (latitude 30°.) it is scarcely habitable, excepting in the valleys of Guasco, Copiapó, and Paposa, which owe their entire fertility to the system of irrigation, invented by the aboriginal Indians and followed by the Spanish colonists. Northward of these places, the absolute desert of Atacama forms a complete barrier, and eastward, the snow-clad chain of the Cordillera separates the Zoological province of Chile, from that of the wide plains which extend on the other side of the Andes.

The last district which it is at all necessary for me to mention here, is that of the GALAPAGOS ARCHIPELAGO, situated under the Equator, and between five and six hundred miles West of the coast of America. These islands are entirely volcanic in their composition ; and on two of them the volcanic forces have within late years been seen in activity. There are five principal islands, and several smaller ones : they cover a space of 2° 10′ in latitude, and 2° 35′ in longitude. The climate, for an equatorial region, is far from being excessively hot : it is extremely dry ; and although the sky is often clouded, rain seldom falls, excepting during one short season, and then its quantity is variable. Hence, in the lower part of these islands, even the more ancient streams of lava (the recent ones still remaining naked and glossy) are clothed only with thin and nearly leafless bushes. At an elevation of 1200 feet, and upwards, the land receives the moisture condensed from the clouds, which are drifted by the trade wind over this part of the ocean at an inconsiderable height. In consequence of this, the upper and central part of each island supports a green and thriving vegetation ; but from some cause, not very easily explained, it is much less frequented, than the lower and rocky districts are, by the feathered inhabitants of this archipelago.

By a reference to the localities here described, it is hoped that the reader will obtain some general idea of the nature of the different countries inhabited by the several animals, which will be described in the following sheets.

The vertebrate animals in my collection have been presented to the following museums : — the Mammalia and Birds to the Zoological Society; the Fishes to the Cambridge Philosophical Society; and the Reptiles, when described, will be deposited in the British Museum. For the care and preservation of all these and other specimens, during the long interval of time between their arrival in this country and my return, I am deeply indebted to the kindness of the Rev. Professor Henslow of Cambridge. With respect to the gentlemen, who have undertaken the several departments of this publication, I hope they will permit me here to express the great personal obligation which I feel towards them, and likewise my admiration at the disinterested zeal which has induced them thus to bestow their time and talents for the good of Science.

MAMMALIA.

Family—PHYLLOSTOMIDÆ.

Desmodus D'Orbignyi.

Plate I. Natural size. Skull, teeth, &c. Pl. XXXV., figs. 1.

D. pilis nitidis adpressis; corpore suprà fusco, pilis ad basin albis; gulâ abdomineque cinerescenti-albis; nasûs prosthemate parvulo bifido.

Description.—The fur of this Bat is glossy and has a silk-like appearance; that on the top of the head, sides of the face, and the whole of the upper parts of the body, is of a deep brown colour; all the hairs on these parts, however, are white at the base. The flanks, interfemoral membrane, and the arms, are also covered on their upper side with brown hairs. On the lower part of the sides of the face, and the whole of the under parts of the body, the hairs are of an ashy-white colour. The membrane of the wing is brownish. The ears are of moderate size, and somewhat pointed; externally they are covered with minute brown hairs, and internally with white. The tragus is also covered with white hairs; it is of a narrow form, pointed at the tip, and has a small acute process in the middle of the outer margin. The nose-leaf is pierced by the nostrils, which diverge posteriorly, and is so deeply cleft on its hinder margin, that it may be compared to two small leaflets joined side by side near their bases. These leaflets, unlike the nose-leaf of the Phyllostomina, lie horizontally on the nose to which they are attached throughout, a slight ridge only indicating their margin. Around the posterior part of the nose-leaf there is a considerable naked space, in which two small hollows are observable, situated one on each side, and close to the

D

nose-leaf; and, at a short distance behind the nose-leaf, this naked membrane is slightly elevated, and forms a transverse fleshy tubercle.

		In.	Lines.						In.	Lines.
Length of head and body	.	3	3	Length of tarsus (claw included)	.	.	0	8½
interfemoral membrane	.	0	3½	ear	.	.	.	.	0	4
the antibrachium	.	2	2	tragus	.	.	.	.	0	3
thumb (claw included)	.	0	8	nose-leaf	.	.	.	0	2¼
tibia	.	0	10	Expanse of the wings	.	.	.	12	8

Habitat, Coquimbo, Chile. (*May.*)

"The Vampire Bat," says Mr. Darwin in his MS. notes upon the present
species, "is often the cause of much trouble, by biting the horses on their
withers. The injury is generally not so much owing to the loss of blood, as to
the inflammation which the pressure of the saddle afterwards produces. The
whole circumstance has lately been doubted in England; I was therefore for-
tunate in being present when one was actually caught on a horse's back. We
were bivouacking late one evening near Coquimbo, in Chile, when my servant,
noticing that one of the horses was very restive, went to see what was the matter,
and fancying he could distinguish something, suddenly put his hand on the beast's
withers, and secured the Vampire. In the morning, the spot where the bite had
been inflicted was easily distinguished from being slightly swollen and bloody.
The third day afterwards we rode the horse, without any ill effects.

Before the introduction of the domesticated quadrupeds, this Vampire Bat
probably preyed on the guanaco, or vicugna, for these, together with the puma,
and man, were the only terrestrial mammalia of large size, which formerly inhabited
the northern part of Chile. This species must be unknown, or very uncommon in
Central Chile, since Molina, who lived in that part, says (Compendio de la Historia del Reyno de Chile, vol. i. p. 301,) "that no blood-sucking species is found
in this province."

It is interesting to find that the structure of this animal is in perfect accordance with the habits as above detailed by Mr. Darwin. Among other points, the
total absence of true molars, and consequent want of the power of masticating
food, is the most remarkable. On the other hand we find the canines and incisors perfectly fitted for inflicting a wound such as described, while the small
size of the interfemoral membrane (giving freedom to the motions of the legs,)
together with the unusually large size of the thumb and claw, would enable
this Bat, as I should imagine, to fix itself with great security to the body of the
horse.

I have named this species after M. d'Orbigny, who has added so much to

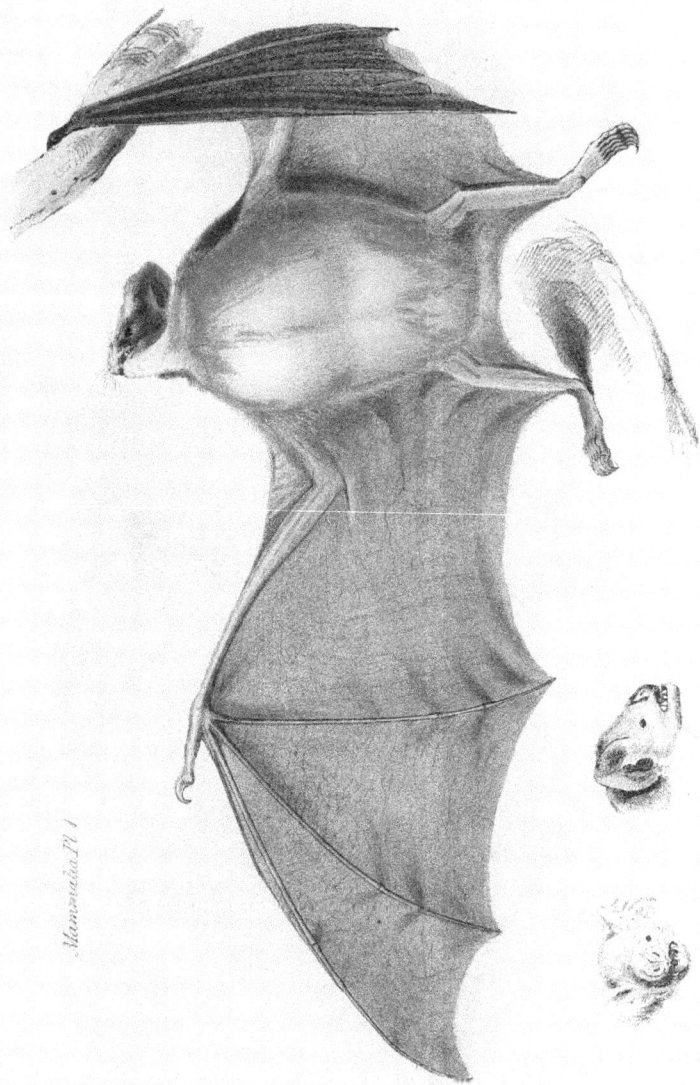

Rhinolophus D'Orbigny

our information on the zoological productions of South America. The *Edostoma cinerea** of that author has evidently a close affinity to the animal here described, and differs chiefly (judging from the drawing published in his work) in the larger size of the ears, in having the nose-leaf free, and the surrounding membrane free and elevated.

As M. d'Orbigny has not yet published the character of his genus *Edostoma*, his figure is my only guide, and in this figure I find the dentition agreeing both with that of the present species, and that of the genus *Desmodus* of Prince Maximilian,—as would appear from the published descriptions, and figure given by M. de Blainville†.—The points of distinction between M. d'Orbigny's animal and the species here described, are not, in my opinion, of sufficient importance to constitute generic characters, I have, therefore, retained the name of Desmodus.

It is desireable perhaps to separate the Blood-sucking Bats from the Insectivorous species, and place them between the latter group and the *Pteropina*, (with which they agree in the large size of the thumb and the rudimentary interfemoral membrane,) under a sectional name, which I propose to call *Hæmatophilini*.

1. PHYLLOSTOMA GRAYI.

PLATE II.

P. fusco-cinereum ; nasús prosthemate lanceolato ; auribus mediocribus, trago basin versus extùs unidentato ; caudá gracillimá, brevi, et membraná interfemorali inclusá ; verrucá complanatá ad apicem menti, verrucis parvulis circumdatá.

DESCRIPTION.—This Phyllostoma agrees with the species described by Mr. J. Gray‡ under the name of Childreni, in having on the lower lip " an half ovate group of crowded warts," but is of a much smaller size, and differs also in colour.

The number of teeth are as follows:—incisors $\frac{4}{4}$; canines $\frac{1}{1}$; molars $\frac{5-5}{5-5}$=32. The intermediate pair of incisors of the upper jaw are large, compressed, and have their apices rounded ; the lateral pair are so minute, that they are scarcely visible without the assistance of a lens : the four incisors of the

---

* Voy. Amer. Merid. t. 8.

† See his memoir " Sur quelques anomalies du système dentaire dans les mammifères," published in the " Annales Françaises et Etrangères d'Anatomie et de Physiologie," No. 6, pl. IX. fig. 2.

‡ Magazine of Zoology and Botany, No. 12.

lower jaw, are somewhat crowded, the intermediate pair are slightly larger than the lateral ; they are all deeply notched, and broad at the apex.   The cerebral portion of the skull is much arched and the anterior portion is depressed.   The zygomatic arch is imperfect; see Pl. 35. figs. 2.   The nose-leaf is lanceolate, and of moderate size : the ears are also of moderate size ; they are rounded at the tip and emarginated on their exterior edge : the tragus is elongated, and suddenly attenuated towards the apex ; the outer margin is deeply notched towards the base, and very obscurely crenulated above this notch.   The interfemoral membrane is of moderate extent, and emarginated posteriorly.   The tail, which is very slender, is entirely enclosed by the interfemoral membrane, and the visible portion appears to consist of but two joints, which together, measure about two and a half lines in length. The basal half of the thumb is enclosed in membrane.   The fur is soft and rather long.   The general tint of the upper and under parts of the body is brownish-ash ; the hairs on the neck and on the whole of the back are grey at the base, then white, or nearly so, brownish-ash near the tip, and whitish at the tip.   On the belly the hairs are nearly of an uniform brown-ash colour, their apices only being whitish.   The ears, nose-leaf, and membrane of the wings, are of a sooty-black hue.

					In.	Lines.						In.	Lines.
Length of head and body	.	.	.	.	2	0	Length of ear	.	.	.	.	0	7
antibrachium	.	.	.	.	1	4⅓	nose-leaf	.	.	.	.	0	3⅓
thumb (claw included)	.	.	.	0	5½	Expanse of the wings	.	:	.	.	10	0
tibia	.	.	.	.	0	7							

Habitat, Pernambuco, Brazil. (*August.*)

" This species appeared to be common at Pernambuco (five degrees north of Bahia).   Upon entering an old lime-kiln in the middle of the day, I disturbed a considerable number of them : they did not seem to be much incommoded by the light, and their habitation was much less dark than that usually frequented as a sleeping place by these animals." D.

I have named this species after Mr. John Gray, the author of several extensive memoirs on the order to which it belongs, and to whom I am indebted for valuable assistance whilst comparing this and other species with those contained in the collection of the British Museum.

2. PHYLLOSTOMA PERSPICILLATUM.

I find in Mr. Darwin's collection, a bat agreeing with the description of M.

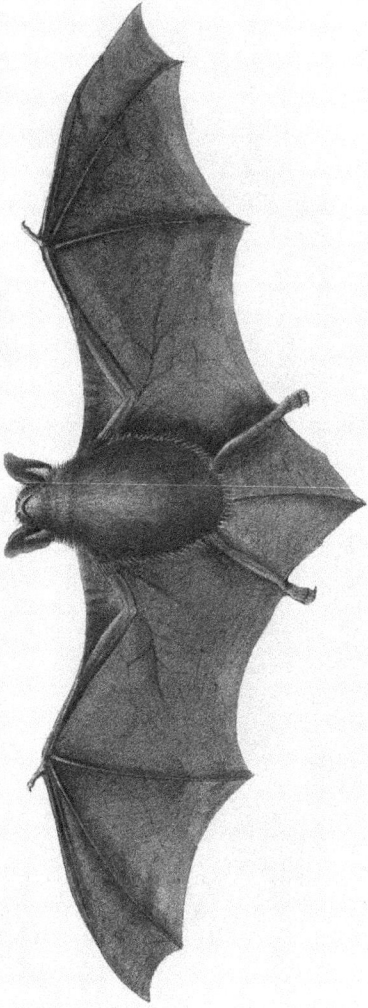

Vespertilio Chilensis

Geoffroy Saint Hilaire,* under the above name, with the exception of a slight difference in the dimensions ; I will, therefore, add those of the present specimen, which is a female. It may be observed, that in the animal before me, the tragus of the ear is pointed, and not bifid at the apex, as represented in plate xi of the work quoted.

	In.	Lines.		In.	Lines.
Length of head and body	4	0	Length of tragus	0	3
antibrachium	2	7	tibia	1	0
nose-leaf	0	5	Expansion of the wings	16	8
ear	0	8½			

" This bat was caught at Bahia, (latitude 13° S.) on the coast of Brazil, in consequence of its having flown into a room where there was a light. I scarcely ever saw an animal so tenacious of life." D.

Family—VESPERTILIONIDÆ.

Vespertilio Chiloensis.

Plate III.

V. fuscus : auribus mediocribus ; trago elongato, angusto, apicem versus attenuato ; fronte concavo ; rostro obtuso ; caudâ ad apicem extremum liberâ.

Description.—In size and colouring, this Bat very closely resembles the *Vespertilio Pipistrellus* of Europe ; the wings, however, are considerably broader in proportion ; the antibrachium, tibia, and tail, are each of them longer ; the tragus of the ear is also longer, and narrower.

The muzzle is short and obtuse, and furnished on each side with numerous hairs, which, when compared with those of other parts, are of a more harsh nature. The nose is naked at the apex. The forehead is concave. The ears are narrow, and somewhat pointed, emarginated externally, and have about four transverse rugæ : the tragus is elongated, narrow, and pointed, and has the outer margin very obscurely crenulated. On the chin there is a small wart, from which spring several stiffish hairs. The tail is about equal to the body in length, and has the extreme tip free. The fur is moderately

* " Annales des Muséum d'Histoire Naturelle," tom. xv. p. 176.

long, and of an uniform rich brown colour, and extends on to the base of the
interfemoral membrane above and below ; the remainder of this membrane is
bare, and, together with that of the wings, of a black colour.

	In.	Lines.		In.	Lines.
Length of the head and body	1	8	Length of the tragus	0	3⅓
the tail	1	3½	the antibrachium	1	5½
Expanse of the wings	8	3	the thumb (claw included)	0	2½
Length of the ear	0	5½	the tibia	0	6¾

Habitat, Chiloe. (*January.*)

" This specimen was given me by Lieut. Sulivan, who obtained it amongst
the islets on the Eastern side of Chiloe. It is not, I believe, common, nor do the
humid and impervious forests of that island appear a congenial habitation for
members of this family. It must, however, be observed, that even in Tierra del
Fuego, where the climate is still less hospitable, and where the number of
insects is surprisingly small, I saw one of these animals on the wing." D.

FAMILY—NOCTILIONIDÆ.

DYSOPES NASUTUS.

Molossus nasutus *Spix*, Simiarum et Vespertilionum. Braziliensium species novæ. Nyctinomus Braziliensis.—*Geoffroy*, Annales des Sciences Naturelles, tom. i. p. 337. pl. 22.

OF this species I find three specimens in Mr. Darwin's collection — " It is remarkable," says Mr. Darwin, " for its wide geographical range. I obtained
specimens at Maldonado, on the northern bank of the Plata, where it was exceedingly numerous in the attics of old houses, and likewise at Valparaiso in
Chile. Molina (vol. i. p. 301.) says another species is found in Chile, of the
same size and figure, but of a more orange (*naranjado*) colour."

Upon comparing the dimensions of several specimens of this species with
those given by Temminck in his " Monographie sur le Genre Molosse," I find
that they vary very considerably ; I shall therefore be adding some little to the
history of the species, by giving the dimensions of those now before me, together
with the sexes of the specimens measured, and their localities. In all these
specimens there is a series of pointed tubercles along the upper margin of the
ears, a character which M. Temminck has omitted to notice. They vary slightly

in the intensity of their colouring, but among those brought from Chile I do not perceive any agreeing with that species, or variety, mentioned by Molina as approaching to an orange colour. All the specimens whose dimensions are here given, are preserved in spirit. Two of them are from Maldonado brought by Mr. Darwin; three were collected in Hayti by Mr. J. Hearne, and one is from Chile, whence it was brought by Mr. H. Cuming.

	From Chile.		Hayti.		Hayti.		Hayti.		Maldonado.		Maldonado.	
	♀		♀		♂		♂		♀		♀	
	In.	Lines.	In.	Lines.	In.	Lines.	In.	Lines.	In.	Lines.	In.	Lines.
Length of head and body . .	2	3	1	11	2	0	2	0½	2	6	2	6
of tail	1	1½	1	2	1	2	1	1½	1	1	1	2
of free portion of ditto .	0	6½	0	5	0	6½	0	5⅓	0	8¾	0	8½
Expanse of wings . . .	10	3	9	3	9	8	9	0	10	6	10	2
Length of antibrachium . .	1	7	1	6	1	6½	1	6	1	8	1	9
of ears . . .	0	5	0	4½	0	4¾	0	4½	0	5⅓	0	5½
Width of ditto . . .	0	7	0	6	0	6	0	6	0	7	0	7
Length from nose to eye . .	0	3½	0	3	0	3¾	0	3	0	3½	0	3½

In all the specimens examined by me, there are two incisors in the upper jaw, and four in the lower, they would therefore, according to M. Temminck, be adult.

Family—CARNIVORA.

1. Canis Antarcticus.

Plate IV.

Antarctic Wolf, *Pennant*, History of Quadrupeds, vol. i. p. 257. sp. 165.
Canis Antarcticus, *Shaw*, Gen. Zool. vol. i. pt. 2. p. 331.
————, *Desm.* Mamm. p. 199.

C. *suprà sordidè fulvescenti-brunneus, pilis ad apicem nigris; lateribus, corporeque subtùs, sordidè flavescenti-fuscis; capite, auribusque extùs, fusco nigroque adspersis; artubus flavescenti-fulvis; labiis, gulâ, abdomine imo, femoribusque intùs, sordidè albis; caudâ ad basin concolore cum corpore, deìn nigrâ, apice albo.*

Description.—This animal is considerably larger than the common fox, (*Canis Vulpes*, Auct.) and stouter in its proportions, and, in fact, appears to be intermediate between the ordinary foxes and the wolves. The tail is much smaller and less bushy than in the former animals. The contour of the head is wolf-like ; the legs, however, are shorter than in the true wolves ; and the tail is white at the apex, a character common in the foxes.

The fur of the Antarctic Fox is moderately long, and the under fur is not very abundant, especially as compared with that of the C. *magellanicus*. This under fur is of a pale brown colour ; the apical portion of each hair is yellowish ; the longer hairs are black at the apex, brown at the base, and annulated with white towards the apex. In many of these hairs the subapical pale ring is wanting. On the chest and belly the hairs are of a pale dirty yellow colour, gray-white at the base, and black at the apex. On the hinder part of the belly the hairs are almost of an uniform dirty white. The space around the angle of the mouth, the upper lip, and the whole of the throat, are white. The chin is brown-white, or brownish. The basal half of the tail is of the same colour as the body, and the hairs are of the same texture ; on the apical half of the tail they are of a harsher or less woolly nature, of a black colour at the apex, and brownish at the base ; those at the extreme point are totally white. The legs are almost of an uniform fulvous colour ; the feet are of a somewhat paler hue ; the hairs on the under side of the hinder feet are brownish, and the external and posterior parts of the tibiæ are suffused with the same tint. The hairs on the head are grizzled with black and fulvous ; the former of these colours is somewhat conspicuous, excepting in the region of the eyes, where the fulvous or yellowish tint prevails. The muzzle is scarcely of so dark a hue as the crown of the head. The ears are furnished internally with long white hairs, externally the hairs are yellowish, with their apices black ; the latter colour is more conspicuous towards the tip of the ear. The sides of the neck near the ear are of a rich fulvous hue.

	In.	Lines.		In.	Lines.
Length from nose to root of tail . .	36	0	Length of ear	2	9
from tip of nose to ear . .	7	3	Height of body at shoulders . .	15	0
of tail (hair included) . .	13	0			

Habitat, Falkland Islands.

"Three specimens of this animal were brought to England by Capt. FitzRoy ; from one of which, the above drawing and description has been made. The earliest notice I can find of this animal is by Pernety,* during Bougainville's voyage, which was undertaken in 1764, for the purpose of colonizing these islands. The strange familiarity of its manner seems to have excited the fears of some of

* Journal Historique d'un Voyage fait aux Iles Malouines, tom. ii. p. 459.

Canis endaurocitus

the seamen in Commodore Byron's voyage (in 1765) in rather a ludicrous manner. Byron says that seals were not the only dangerous animals that they found, " for the master having been sent out one day to sound the coast upon the south shore, reported at his return that four creatures of great fierceness, resembling wolves, ran up to their bellies in the water to attack the people in his boat, and that as they happened to have no fire-arms with them, they had immediately put the boat off in deep water." Byron adds that, " When any of these creatures got sight of our people, though at ever so great a distance, they ran directly at them ; and no less than five of them were killed this day. They were always called wolves by the ship's company, but, except in their size, and the shape of the tail, I think they bore a greater resemblance to a fox. They are as big as a middle-sized mastiff, and their fangs are remarkably long and sharp. There are great numbers of them upon this coast, though it is not perhaps easy to guess how they first came hither; for these islands are at least one hundred leagues distant from the main. They burrow in the ground like a fox, and we have frequently seen pieces of seals which they have mangled, and the skins of penguins lie scattered about the mouths of their holes. To get rid of these creatures, our people set fire to the grass, so that the country was in a blaze as far as the eye could reach, for several days, and we could see them running in great numbers to seek other quarters."

The habits of these animals remain nearly the same to the present day, although their numbers have been greatly decreased by the singular facility with which they are destroyed. I was assured by several of the Spanish countrymen, who are employed in hunting the cattle which have run wild on these islands, that they have repeatedly killed them by means of a knife held in one hand, and a piece of meat to tempt them to approach, in the other. They range over the whole island, but perhaps are most numerous near the coast ; in the inland parts they must subsist almost exclusively on the upland geese, (*Anser leucopterus,*) which, from fear of them, like the eider-ducks of Iceland, build only on the small outlying islets. These wolves do not go in packs ; they wander about by day, but more commonly in the evening ; they burrow holes; are generally very silent, excepting during the breeding season, when they utter cries, which were described to me as resembling those of the *Canis Azaræ.* Spaniards and half-cast Indians, from several districts of the southern portions of South America, have visited these islands, and they all declare that the wolf is not found on the mainland ; the sealers likewise say it does not occur on Georgia, Sandwich Land, or the other islands in the Antarctic ocean. I entertain, therefore, no doubt, that the *Canis antarcticus* is peculiar to this archipelago. It is found both on East and West Falkland, as might have been inferred from the accounts given by Bougainville and Byron, who visited different islands ;—I state this particularly, because the contrary has been asserted. I was

c

assured by Mr. Low, an intelligent sealer, who has long frequented these islands, that the wolves of West Falkland are invariably smaller and of a redder colour than those from the Eastern island; and this account was corroborated by the officers of the Adventure, employed in surveying the archipelago. Mr. Gray, of the British Museum, had the kindness to compare in my presence the specimens deposited there by Captain Fitzroy, but he could not detect any essential difference between them. The number of these animals during the last fifty years must have been greatly reduced; already they are entirely banished from that half of East Falkland which lies East of the head of St. Salvador Bay and Berkeley Sound; and it cannot, I think, be doubted, that as these islands are now becoming colonized, before the paper is decayed on which this animal has been figured, it will be ranked amongst those species which have perished from the face of the earth."—D.

2. Canis Magellanicus.

Plate V.

Canis Magellanicus, *Gray*, Proceedings of the Zoological Society of London, part iv. 1836, p. 88.
Vulpes Magellanica, *Gray*, Magazine of Natural History, New Series, 1837, vol. i. p. 578.

C. *suprà albo nigroque variegatus; lateribus fulvescente fuscoque lavatis; capite fusco-flavo et albescente adsperso; rostro supernè obscuriore; auribus, artubusque extùs flavescenti-rufis; corpore subtùs sordidè flavescenti-albo; pectore fulvo lavato; mento fuscescente; caudâ fulvescenti-fuscâ, pilis ad apicem nigris, subtùs pallidiore; plagâ supernè prope basin caudæ, hujusque apice nigris.*

Description.—This species is considerably larger than the European fox; its form is more bulky, the limbs are shorter and stouter in proportion, the ears are smaller and the tail is more bushy. The fur is long, thick, and loose. The under fur is very long, abundant, and of a woolly texture. The back is mottled with black and white, the former of these colours being predominant; the hairs on this part are gray at the base, there is then a considerable space of a pale, or whitish brown colour; next follows a broad white ring, beyond which the hairs are black. On the sides of the body the hairs are coloured in the same way, excepting that the white portion is more extended, and is followed by a rich yellow-brown, shaded into black as it approaches the apex of each hair. Hence the general hue of the sides of the body is paler than that of the back, the brown and white tints being the more conspicuous.

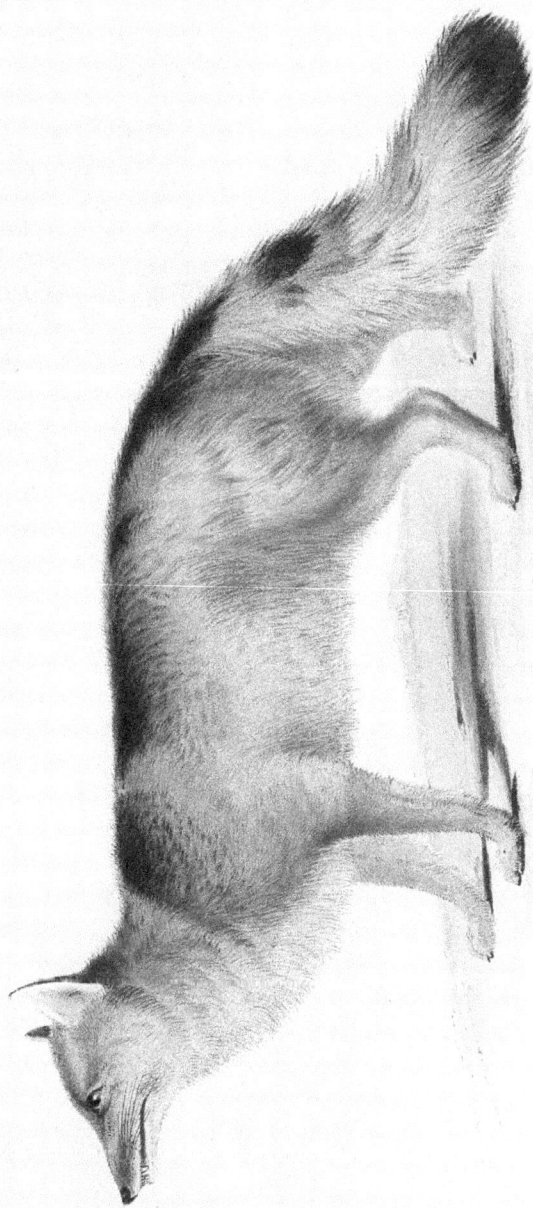

The hairs of the head are annulated with white, and fulvous, and are black at the tip ; the two former colours are most conspicuous. The chin is brownish. The lower part of the cheeks, the throat, and the under parts of the body, are of a dirty yellowish white colour, inclining to buff in certain parts, especially on the lower part of the neck and chest. The limbs are of a rich deep fulvous, or yellowish rust colour externally ; the feet and inner sides of the legs are of a paler hue. On the hinder legs externally, above the heel, is a patch of bright rust colour ; such is also the colour of the ears externally, and likewise of that portion of the neck behind the ears. Internally the ears are furnished with long yellowish white hairs. The tail is long and very bushy; at its base the hairs are rusty white, towards the middle they are of a paleish rust colour, and at the apex they are black; there is also a black patch on the upper part towards the base. The hairs of the tail beneath are almost entirely of an uniform rusty white colour, those on the upper side are all tipped with black.

	In.	Lines.		In.	Lines.
Length from nose to root of tail .	. 31	0	Length of ear	. 2	0
to base of ear .	. 6	9	Height of body at the shoulders .	. 14	6
of tail (hair included) .	. 17	0			

Habitat, Chile. (*June.*)

" This animal was first brought to Europe by Captain Philip P. King, who obtained it at Port Famine in Tierra del Fuego, where it is common. My specimen was obtained in the valley of Copiapó in the northern part of Chile. The Magellanic fox, therefore, has a range on the western coast of at least 1600 miles, from the humid and entangled forests of Tierra del Fuego, to the almost absolutely desert country of northern Chile. In La Plata, on the Atlantic side of the continent, I believe it is not found.* It is mentioned by Molina in his account of the animals of Chile,† under the name of Culpeu, which he supposes to be derived from the Indian word " culpem," signifying madness ; for this animal, when it sees a man, runs towards him, and standing at the distance of a few yards, looks at him attentively. He adds, although great numbers are killed, they do not leave off this habit. Molina states that he has repeatedly been a witness of this, and I received nearly similar accounts from several of the inhabitants of Chile : yet I must observe, that the people of the farm-house, where my specimen was killed

---

* Azara has not described this animal, which circumstance alone would render it probable that it is not an inhabitant of Paraguay or La Plata. The two Foxes mentioned by him are the Aguará-guazú, (*Canis jubatus*, Auct.) a very large kind of fox (a strangely exaggerated description of this animal is given by Falkner) of which I could not obtain a specimen ; and the Aguará-chay, or *Canis Azaræ.*

† Molina, Compendio de la Historia del Reyno de Chile, vol. i. p. 330 and 332.

(after it, together with its female, had destroyed nearly two hundred fowls) bitterly complained of its craftiness. From this bold curiosity in the disposition of the Culpeu, Molina thought that it was the same animal as that described by Byron at the Falkland Islands, but we now know that they are different. The Culpeu burrows holes under ground, often wanders about by day, is very strong and fleet. When riding one day in the valley of Copiapó, accompanied by a half-bred grey-hound, I happened to come across one of these foxes ; and although the ground was, in the first part of the chase, level, it soon entirely distanced its pursuer. Whilst running, it barked so like a dog, that until it had run some way a-head of the greyhound, I could not tell from which animal the noise proceeded. After the Culpeu had reached the mountains, it made a sudden bend from its course, and returned in a nearly parallel line, but at the base of a steep cliff of rock ; it then quietly seated itself on its haunches, and seemed to listen with much satis-faction to the dog, which was running the scent on the mountain side, above its head."—D.

3. CANIS FULVIPES.

PLATE VI.

Canis fulvipes, *Martin*, Proceedings of the Zoological Society of London, 1837, p. 11.

C. *suprà niger, albo adspersus, capite lateribusque fuscis, sordidè albo nigroque adspersis ; rostro superiore, mentoque fusco-nigricantibus ; gulá, labiis superioribus, femoribusque ad partem anteriorem, sordidè albis ; pectore abdomineque fuscescentibus ; auribus externè rufo-castaneis ; brachiis internè, tarsis, digitisque fuscescenti-fulvis ; artubus posticis extùs supra calcem fusco-nigrescentibus ; caudæ colore ad basin ut in corpore, apice nigro.*

DESCRIPTION.—This species is considerably less than the common European fox, (*Canis Vulpes*, Auct.) its weight probably would scarcely exceed half that of the latter animal. The form of the body is stout, the limbs are short and rather slender ; the head is also short, and the muzzle is pointed ; the ears are of moderate size. The tail is about equal to half the whole length of the body, head included ; and compared with that of ordinary foxes, is much less bushy, especially at the base. The general hue of this animal is very dark ; the fur is rather short, and harsh to the touch ; the under fur is abundant, and of a woolly texture. On the back, all the hairs are of a deep brown colour, annulated with white near the apex, and black at the apex. When the fur is

in its ordinary position, the brown colour is not seen, and the black and white produce a grizzled appearance; the black colour, however, predominates. On the sides of the body each hair is grayish at the base, then pale brown, near the apex annulated with white, and at the apex black: the three last mentioned colours are exhibited in about equal proportions (the fur being in its natural position) over the haunches and shoulders, but between these two parts, the brown and white colours are the more conspicuous. The hairs of the head are coloured in the same way as those of the sides of the body, excepting that the brown portion of each hair, is replaced by rusty brown, which gives a rufous hue to this part. The muzzle and chin are of a sooty brown colour. A dirty white patch is observable on each side of the muzzle at the apex, and this colour is extended along the margin of the upper lip on to the lower part of the cheeks, and over the whole of the throat; all the hairs in these parts (with the exception of those on the lips) being of a deep brownish gray colour, with their apical portions only, white. The ears are covered internally with long yellowish white hairs; towards, and on the margin of the ears externally, the hairs are of a buff colour, on the remaining portion of the ears, and on the sides of the neck, they are of a reddish chestnut hue. The hairs of the under parts of the body are brown, those near the hinder legs, and between them, are of a dirty white colour at the apex; towards the rump they are of a yellowish brown colour. The hairs of the tail are brown, black at the apex, and annulated with white near the apex; on the apical portion the hairs are black, and brown at the base. The fore legs are of a brown colour externally, internally they are of a brownish fulvous hue; such is also the colour of the feet. The fore part of the posterior legs is whitish, and there is a large blackish patch on the outer side, and extending around the posterior part, above the heel.

	In.	Lines.		In.	Lines.
Length from nose to root of tail . .	24	0	Length of ear	2	3½
to base of ear . .	0	4¾	Height of body at shoulders . .	10	6
of tail (hair included) . .	10	0			

Habitat, Chiloe. (*December.*)

" I killed this animal on the sea-beach, at the southern point of the island; it is considered extremely rare in the northern and inhabited districts. Molina mentions this fox, which he falsely considered as the *C. lagopus*, under the name of the *Payne Gurú*, and he adds, that in the Archipelago of Chiloe, it is found of a black colour. From this circumstance I am induced to believe that the species is confined to these islands."—D.

4. Canis Azaræ.

Plate VII.

Canis Azaræ, *Pr. Maximilian*, Beiträge zur Naturgeschichte Braziliens, vol. ii. p. 338.

Agouarachay, *Azara*, Essais sur l'histoire naturelle des Quadrupèdes de la Province du Paraguay, tom. i. p. 317.

C. *suprà albo nigroque variegatus ; lateribus cinerescentibus ; capite, auribus externè, artubusque, cinereo-cinnamominis; mento nigro ; tibiis externis ad basin nigro lavatis ; caudâ albescente, suprà nigro variegatâ, ad apicem nigrâ; spatio pone angulos oris, gutture, corporeque subtùs albescentibus ; fasciis duabus griscescentibus in pectore plus minusve distinctis.*

DESCRIPTION.—Compared with the common fox (*Canis Vulpes*, Auct.), the present animal is rather smaller, and of a more slender form. Its limbs are a little longer in proportion; the ears are not so broad. The tail is not quite so bushy, neither is it so long; the fur is much longer, and of a harsher nature.

The predominant colours of the body are black and white; the limbs are of a fulvous hue externally. The hairs on the under part of the feet are dirty brown; the fore part of the anterior legs, and the feet, are of a buff colour; on the former, the hairs are more or less distinctly tipped with black, which produces a grizzled appearance. The inner side of the fore legs is of an uniform pale buff colour; the hinder part of these legs, the fore part of the posterior legs, and the inner side of the thighs, are white. On the outer side of the hinder legs, at some little distance above the heel, is a large blackish patch. The under parts of the body are of a dirty white hue, arising from the hairs being dusky or brownish at the base, and tipped with white, as on the fore part of the belly, or of a pale buff colour at the base, as towards the rump. The edge of the upper lip, the throat, neck, and chest, are white; a broad grayish band extends across the latter, and another of a paler hue crosses the lower part of the neck. The chin is black, and this colour is extended backwards around the angle of the mouth. The upper part of the head is of a pale yellow-brown colour, each hair being annulated with white near the apex. The ears are furnished with white hairs internally, and externally they are of a yellowish brown colour, tipped with black; at the base of the ears, and the portion of the neck on each side nearest to them, the

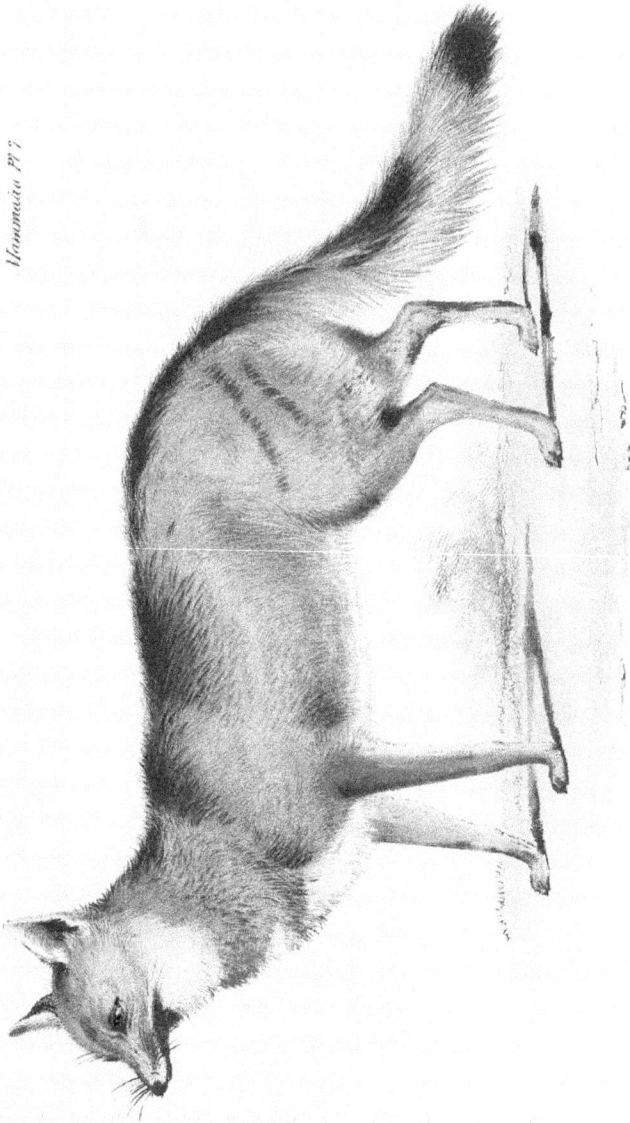

Canis Azara

hairs are of an uniform buff colour. The hairs of the moustaches are long and stiff, and of a black colour. The hairs of the back, which are very long, are brown at the base, very pale towards the skin, and of a deep brown in the opposite direction ; each hair is then white, and at the apex black. The tail is whitish, mottled with black ; the apical portion is black, and there is a patch of the same colour towards the base on the upper side.

	In.	Lines.		In.	Lines.
Length from nose to root of tail . .	27	6	Length of ear	3	2
to base of ear . .	5	9	Height of body at shoulders . .	14	0
of tail (hair included) . .	14	6			

Habitat, La Plata, Patagonia, and Chile.

The black and white portions of the hairs on the back produce in that part a mottled appearance, and in the specimen from which the above description is taken, these two colours are about equal in proportion. In another specimen now before me, the black colour predominates on the back. The fur in the younger animals of this species is not so long nor so harsh, and the upper parts are grizzled with black and white ; that is to say, these two colours do not form patches of considerable extent as in the adults ; the general colouring is also somewhat paler. The chin is brown-black or brown, instead of black, and the upper band, or that, which in the adult extends across the upper part of the neck, is interrupted in the middle ; in fact, is only traceable on the *sides* of the neck.

Azara, in his description of the Agouarachay, says, the muzzle, as far back as the eyes, is blackish ; whereas, in all the specimens examined by me, the muzzle is of the same colour as the other parts of the head, or *very* nearly so. In other respects his description agrees with the animal described by me, and *not* with the *Canis cinereo-argentatus*, which Desmarest and Lesson suppose to be the Agouarachay of Azara. In Fischer's "*Synopsis Mammalium*" the *Canis Azaræ* is described as having the tip of the tail white ; whereas it is black, not only in the five specimens which have come under my notice, but also in those in the collection of Prince Maximilian* (who was the original describer) and in the Paris Museum.

"This animal has a wide range ; Prince Maximilian brought specimens from Brazil ; and it is common in La Plata, Chile, the whole of Patagonia, even to the shores of the Strait of Magellan ; and a fox, which lives on the small islands not far from Cape Horn, probably belongs to this species. This animal generally frequents desert places ; I saw many in the valley of the Despoblado, a branch of

* I am indebted to Mr. Ogilby, who visited the Prince's collection, for a description from the specimens of *C. Azaræ* therein preserved. In this description the tip of the tail is said to be black.

that of Copiapó, where there is no fresh water, and where, with the exception of some small rodents, (the constant inhabitants of sterile regions) scarcely any other animal could exist. I saw also very many of these foxes wandering about by day (although Azara says they are nocturnal in Paraguay) on the plains of Santa Cruz, where various kinds of mice are abundant, and likewise around the Sierra Ventana. In the course of one day's ride in this latter neighbourhood, (not far from Bahia Blanca, lat. 39° S.) I should think I saw between thirty and forty. They generally were wandering at no great distance from their burrows ; but, as they are not very swift animals, our dogs caught two. Azara states that in Paraguay this fox, which he calls the Agourá-chay, inhabits thick woods, and that it makes a great nest or pile of straw, to lie on ; but that near Buenos Ayres it uses the holes of the Bizcacha. Further southward, where the Bizcacha is not found, it certainly excavates its own burrow.* In Chile these foxes are very destructive to the vineyards, from the quantity of grapes they consume ; so that boys are generally kept in the vintage season with bells and other means to frighten them away. Azara states, that in Paraguay they likewise eat fruit and sugar-cane. By the same authority it is said, that the Agourá-chay, when taken young, is easily domesticated."—D.

1. FELIS YAGOUAROUNDI.

PLATE VIII.

Felis Yagouaroundi, *Desmarest*, Mammologie, p. 230.

Yagouaroundi, *Azara*, Essais sur l'histoire Naturelle des Quadrupèdes de la Province du Paraguay, tom. i. p. 171.

Felis Darwinii, *Martin*, Proceedings of the Zoological Society of London, 1837, p. 3.

F. *vellere brevi, adpresso, purpurascenti-fusco ; pilis flavescente annulatis ; pedibus nigro lavatis ; caudá longissimá ; auribus parvulis.*

DESCRIPTION.—The fur is rather harsh, short, and somewhat adpressed : the under fur is of a pale grayish brown colour ; the hairs which constitute the chief clothing of the animal, are black, annulated with brownish yellow, or in some parts, yellow-white, each hair having about three or four rings. The black and pale colours are about equal in proportion, and their mixture pro-

---

* Considering the great difference of climate and other conditions between the hot and wooded country of Paraguay, and the desolate plains of Patagonia, one is led to suspect that the *Canis Azaræ* of La Plata and Patagonia, which wanders about by day, and inhabits burrows instead of heaps of straw, may turn out to be a different species from the Agouará-chay of Azara, which is nocturnal in its habits, and lives in thick coverts.

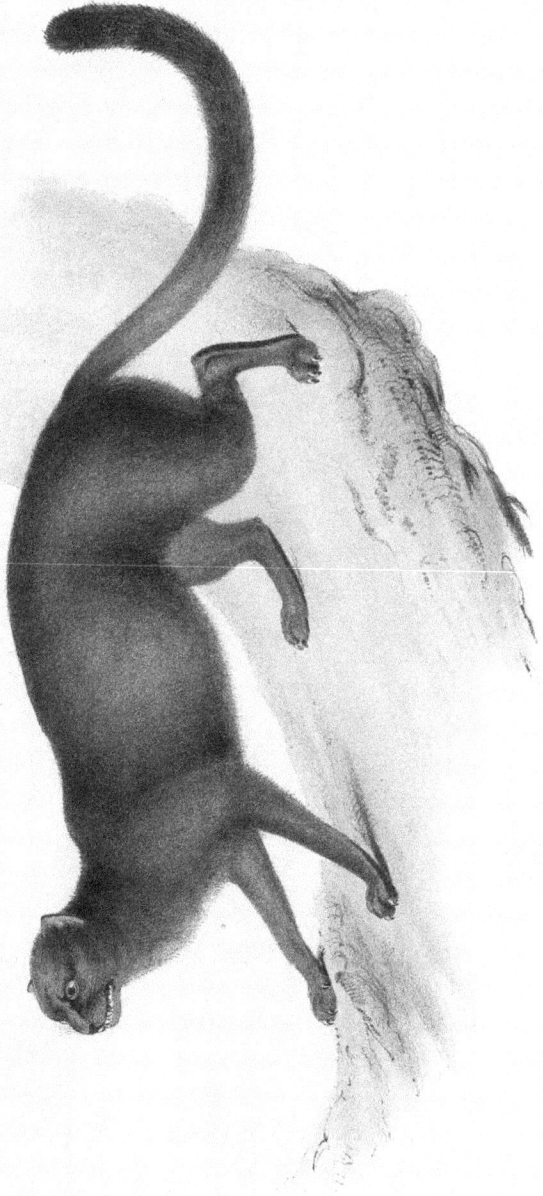

Felis Jaguarondi.

duces a deep brown tint, which is almost uniform throughout the body and limbs. On the head the yellowish colour predominates over the black, excepting on the tip of the muzzle, and thence back to the eye, where the hairs are of a brownish black colour. On the throat the hairs are brown. The underside of the tarsus is black, and on the outer side of the fore-foot there is a black mark which extends upwards on to the wrist. The tail is long and bushy; towards and on the base, the hairs are annulated with black and yellow, like those of the body; but beyond this they are of a more uniform colour, each hair being brown at the base, and gradually shaded into black towards the tip. The ears are small and rounded, and covered with hairs of the same colour as those on the head. The claws are of a large size, and white colour; the toes are united for a considerable portion of their length by the interdigital membrane.

	In.	Lines.		In.	Lines
Length from nose to root of tail .	. 25	0	Length of ear	. 1	0
to base of ear .	. 3	6	Height of body at shoulders	. 12	0
of tail (hairs included) .	. 19	0			

Habitat, Rio de Janeiro, Brazil. (*May.*)

" This cat was given me by an old Portuguese priest, who had hunted it down in a thick forest with a small pack of dogs, after a severe chase. It was killed near the Gavia mountain, at the distance of a few miles only from Rio de Janeiro, where it was considered uncommon." D. Although small, compared with the Puma, (*Felis concolor*, Auct.), this cat, in its slender lengthened body, small head, long tail, and stout limbs, decidedly evinces an affinity to that species. According to the dimensions of the Yagouaroundi given by Azara, Desmarest, and Temminck, it appears that the tail is considerably shorter in proportion in the specimens examined by those naturalists, than in the present individual, and the difference was such, as to induce Mr. Martin to believe that the latter was a distinct species; he accordingly proposed for it the specific name of *Darwinii*. At the time that Mr. Martin described the specimen alluded to, I was also inclined to believe it was a distinct species. I mention this because I am afraid my opinion had a slight share in influencing Mr. Martin's determination. I have since seen many specimens, and upon comparing their dimensions, I find that the proportionate length of the tail varies more than is usual in other species of cats, and that the difference in the length in this member is not combined with any other distinguishing character. In colouring there is also a considerable variation, some specimens being almost black, and having the hairs but obscurely annulated with white; in others, the hairs are more distinctly annulated, and the head assumes a grayish hue. Others again, are brown, or black brown, having the hairs annulated with yellow. The following are the dimensions of two specimens in the Paris museum, and those given by the authors above alluded to.*

	Paris M.		Paris M.		Desmarest.		Temminck.		Azara.	
	In.	Lines.	In.	Lines.	In.	Lines.	In.	Lines.	In.	Lines.
Length from nose to root of tail .	30	6	28	0	23	0	30	0	36	9
of tail	24	0	17	0	13	9	22	0	13	9

2. FELIS PAJEROS.

PLATE IX.

Chat Pampa, *Azara*, Essais sur l'histoire Naturelle des Quadrupèdes du Paraguay. Traduct. Franç. tom. 1. p. 179.

Felis Pajeros, *Desmarest*, Mammologie, p. 231.

F. vellere longissimo, flavescenti-griseo, fasciis flavescenti-fuscis indistinctè et sublon-gitudinalitèr notato; pedibus annulis latis nigris; abdomine maculis magnis nigris; mento albo; caudâ brevi; auribus mediocribus, ad apicem externum nigris.

DESCRIPTION.—The Pampas cat is about equal in size to the common wild cat of Europe (*Felis Catus*, Linn.). It is however of a stouter form than that animal, the head is smaller, and the tail is shorter.

The most remarkable character in this species consists in the great length of the fur, — the longer hairs on the back measuring upwards of three inches, and those on the hinder part of the back, are from four and a half, to four and three quarter inches in length. The general colour of the fur is pale yellow-gray. Numerous irregular yellow, or sometimes brown stripes run in an oblique manner from the back along the sides of the body. On each side of the face there are two stripes of a yellowish or cinnamon colour: these stripes commence near the eye, extend backwards and downwards over the cheeks, on the hinder part of which they join and form a single line, which encircles the lower part of the throat. The tip of the muzzle and the chin are white, and there is a spot in front of the eye, and a line beneath the eye, of the same colour: the belly and the inner side and hinder part of

* In measuring the species of Mammalia, I almost invariably, when wishing to give the length, measure from the tip of the nose *along the curve of the back* to the root of the tail. In the Ruminantia of course this plan is not desirable, but in other Mammals I have found it most convenient. If we take a Cat, for instance, and curve the body in whatever way we please, we find the length (taken in the way just mentioned) always the same. Whereas, if we take a straight line (as many naturalists do) the length will vary according to the position of the animal.

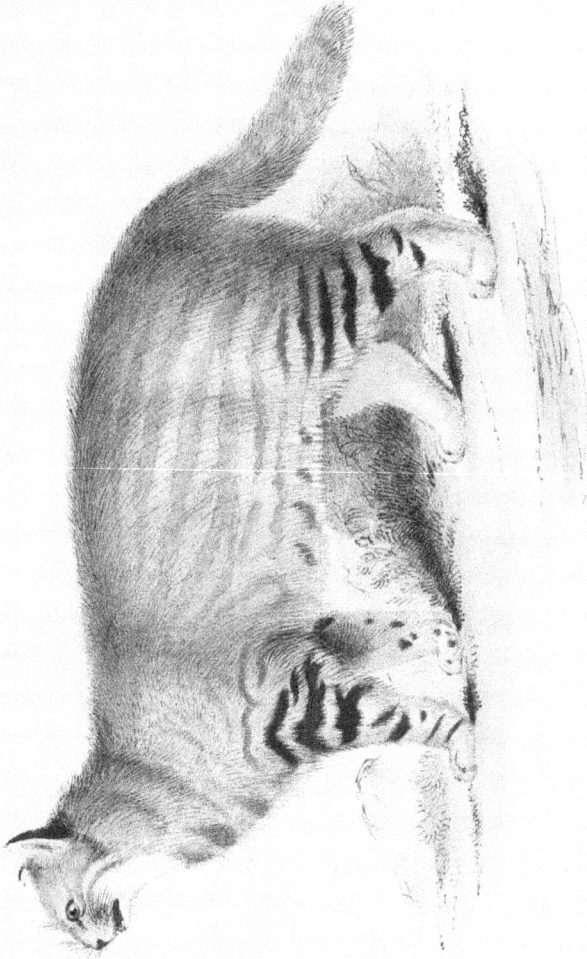

Felis Pajeros

the fore-legs are also white. An irregular black line runs across the lower part of the chest and extends over the base of the fore-legs externally, and above this line there are two other transverse dark markings on the chest, which are more or less defined. On the fore-legs there are three broad black bands, two of which encircle the leg, and on the posterior legs there are about five black bands externally, and some irregular dark spots internally. The feet are yellowish, and the underside of the tarsus is of a slightly deeper hue. On the belly there are numerous large irregular black spots. The ears are of moderate size, furnished internally with long white hairs; externally, the ears are of the same colour as the head, excepting at the apex where the hairs are black and form a slight tuft. The tail is short, somewhat bushy, and devoid of dark rings or spots — the hairs are in fact coloured as those of the back of the animal. On the upper part of the body each hair is brown at the base, then yellow, and at the apex, black. On the hinder part of the back the hairs are almost black at the base, and on the sides of the body each hair is gray at the base; there is then a considerable space of yellowish-white colour; towards the apex they are white, and at the apex black. The greater number of the hairs of the moustaches are white.

	In.	Lines.		In.	Lines.
Length from nose to root of tail .	26	0	Length of ear	1	11
to base of ear .	3	6	Height of body at shoulders .	13	0
of tail (fur included) .	11	0			

Habitat, Santa Cruz, Patagonia, (*April,*) and Bahia Blanca, (*August.*)

The markings in this animal vary slightly in intensity; those on the body are generally indistinct, but the black rings on the legs are always very conspicuous.

"This animal takes its name from 'paja,' the Spanish word for straw, from its habit of frequenting reeds. It is common over the whole of the great plains, which compose the eastern side of the southern part of America. According to Azara, it extends northward as far as latitude 30°, and to the south, I have reason to believe, from the accounts I have received, that it is found near the Strait of Magellan, which would give it a range of nearly 1400 miles, in a north and south line. One of my specimens was obtained, in 50° south, at Santa Cruz: it was met with in a valley, where a few thickets were growing. When disturbed, it did not run away, but drew itself up, and hissed. My other specimen was half-grown, and was killed in the end of August, at Bahia Blanca."—D.

2. Felis domestica.

Felis domestica, *Brisson*, Reg. Anim. p. 264.

I find in Mr. Darwin's collection a cat, the colouring and proportions of which, convince me that its origin is from the domestic cat, as however it was shot in a wild state far from any house, a description may, perhaps, prove useful. Its general colour is deep gray, and the body is adorned with numerous irregular narrow black bands; there is a broad black mark, formed of confluent spots, along the middle of the back, which commences a little behind the shoulders; a considerable space around the angles of the mouth, the chin, throat, central portion of the chest, fore-feet, toes of the hinder feet, and the posterior portion of the belly, are white; a black line extends backwards from the posterior angle of the eye, on to the cheeks; thence, across the throat, there are two lines: the space between the eye is chiefly occupied with white hairs: the tail is slender, and tapers towards the apex; the basal half is gray with black rings, and the apical half is black, excepting the extreme point, which is white: the tarsus is black beneath: the legs are of a deep gray colour, banded with black externally.

To the dimensions I will add those of a domestic cat which in colour and markings very closely resembles the animal above described. I may add that I have chosen a cat rather above the ordinary size for my comparison, yet it will be seen that the wild cat has the advantage in bulk.

	Wild Cat. In. Lines.		Domestic Cat. In. Lines.	
Length from nose to root of tail . . .	22	0	19	0
of tail	12	3	11	6
of tarsus	5	1	4	7
of ear	1	11		
Height at shoulders	11	3		

Habitat, Maldonado, La Plata, (*May.*)

" This animal was killed amongst some thickets on a rocky hill a few miles from Maldonado. It appeared, when dead, much larger and stronger than any domestic cat I ever saw, and it was described to me as having been exceedingly fierce. I mention this because M. Temminck supposes that the domesticated varieties of all animals are of larger size, than the wild stock from which they are descended."—D.*

---

* I must refer the reader to my journal for some account of the habits of the jaguar and puma, which being well known animals, and the facts that I mention having little scientific interest, I have not thought it worth while to repeat them here.

GALLICTIS VITTATA.

Gallictis vittata, *Bell*, Zoological Journal, vol. ii. p. 551–2.
———————, „ Proceedings of the Zoological Society, for April, 1837, p. 39.
Gulo vittatus, *Desmarest*, Mamm. p. 175.

"This animal is not uncommon at Maldonado, where it is called "*Huron*" or thief, from the ravages it commits on eggs and poultry. Shortly after being killed this specimen weighed 1 lb. 8 oz. (Imp. weight)."—D.

1. LUTRA PLATENSIS.

L. vellere nitido, adpresso, intensè fusco; corpore subtùs pallidiore; gutture ad latera, et subtùs, pallidè fusco; mento rostrique apice sordidè flavescenti-albis; pedibus nigrescenti-fuscis; pilis caudæ supernè brevioribus, adpressis, illis ad caudæ latera longioribus et fimbriam efficientibus.

DESCRIPTION.—This Otter is about equal in size to the common European species (*Lutra vulgaris*, Auct.): its fur is short, glossy, and adpressed ; the under fur is tolerably abundant and of a silky nature. The general colouring of the ordinary fur is deep brown, and that of the under fur is very pale brown, deepér externally. The tint of the under parts of the body is paler than the upper, and may be described as brown, that of the throat, sides and under part of the neck, pale brown ; and, on the tip of the muzzle and chin, dirty yellowish-white. The hairs of the moustaches are brownish-white ; the ears are covered with short deep brown hairs, those towards the tip are paler. The hairs covering the feet above are short, and of a very deep brown colour. The tail is tolerably long, thick at the base, whence it gradually tapers to the apex. The hairs on the base of the tail resemble those of the body, but on the remaining portion, they are short, glossy, and very closely applied to the skin both on the upper and under surface, whereas those on the sides are longer, and form a kind of fringe. The tip of the muzzle and the soles of the feet are naked, with the exception of the hinder half of the tarsus.

	In.	Lines.						In.	Lines.
Length from nose to root of tail .	. 28	0	Length of tail .	.	.	.	.	. 18	0

Habitat, La Plata, (*July.*)

The La Plata Otter in its general colouring is of a somewhat deeper hue than the European species, the cheeks and throat instead of being nearly white are of a pale brown colour ; the tail is longer in proportion, and tapers more gradually ; the tip of the muzzle is naked, but the hairless portion is less than in that species, the boundary line between the naked part and the hair of the top of the muzzle forming almost a semicircle ; the retiring extremities of this line touch the posterior angle of the nostril on each side, whereas in the common otter the boundary line of the hair of the muzzle is of a w-like form.   The skull is figured in Plate 35, figs. 4, *a, b, c,* and *d,* and is compared with that of *L. Chilensis* in the next description.

"This specimen was killed by some fishermen a few miles from Maldonado, near the mouth of the estuary of the Plata, where the water is quite salt.   I am not, however, by any means sure that it may not be a fresh-water species, which had wandered from its proper station ; in the same manner as not unfrequently is the case with the *Hydrochærus Capybara.*   I am indebted to Mr. Chaffers, the master of the Beagle, for having kindly presented me with this specimen." — D.

2. LUTRA CHILENSIS.

Lutra Chilensis, *Bennett,* Proceedings of the Committee of Science and Correspondence of the Zoological Society of London for 1832, p. 1.

L. fusca ; vellere mediocri, laxo et sub-extante; mento, gulá, et faciei lateribus, pallidè fuscis ; pedibus saturatè fuscis; corporis pilis ad apicem pallidè fuscis ; caudá mediocri ; rostri apice calvo.

DESCRIPTION.—This species scarcely equals a full grown European otter in size. It is of a brown colour throughout ; the cheeks, chin, and throat, being slightly paler, and the feet of a deeper tint, than the other parts.   The fur is moderately long, rather harsh to the touch, and semi-erect: the under fur is abundant, and of a soft and silk nature.   The hairs of the ordinary fur are deep brown, but tipped with a very pale brown colour.   The hairs of the tail, like those of the body, are harsh and semi-erect; towards the apex, those on the upper and under part are in a slight degree shorter than those at the sides, and lie closer to the skin ; these differences, however, are not very apparent on the upper side, though distinct on the under.   The feet are naked beneath, with the exception of the posterior half of the tarsus. The hair of the muzzle extends only down to the posterior angle of the nostrils, where it terminates in a straight line, leaving the tip of the muzzle naked.

	In.	Lines.			In.	Lines.
Length from nose to root of tail .	. 31	0	Length of tail	14	3

Habitat, Chonos Archipelago, (*January.*)

The Chile Otter was originally described by Mr. Bennett from a specimen presented to the Zoological Society by Mr. Cuming, but as this specimen is a young animal, scarcely half-grown, it does not present some of the characters of the species in so marked a manner as the adult. I have, therefore, availed myself of an adult specimen in Mr. Darwin's collection, to draw up the above description.

Compared with the Common Otter (*Lutra vulgaris*, Auct.) the most striking difference consists in the character of the fur: the hairs instead of being adpressed as in that species, are here semi-erect, and appear as if they had been clipped at the extremity. The fur is of a deeper colour, but has a slightly grizzled appearance, owing to the tip of each hair being of a much paler colour than the remaining part.

In the young animal described by Mr. Bennett, (which in weight was probably not more than one-third of that of the present animal) the hairs of the body are of an uniform deep brown colour; hence, if I am right in considering Mr. Darwin's animal as the same species, it would appear that the grizzled character of the fur is dependent on age.

The semi-erect fur will also serve to distinguish the present species from the *Lutra Platensis;* the fur is likewise longer, the tail is shorter, and the feet are smaller in proportion. The most important distinctions, however, are furnished by the skulls; I will, therefore, compare them.

The skull of *L. Chilensis* compared with that of *L. Platensis*, (Plate 35, figs. 4.) when viewed from above, presents but little difference in general form ; it is, however, smaller in all its proportions, and the zygomatic arch is a little less convex: the palate is proportionately shorter; the tympanic bullæ are much smaller, less elevated, and wider apart, in which respect there is a greater approximation to the skull of *L. vulgaris* than to that of *L. Platensis;* but here, the tympanic bullæ are larger than in *L. Chilensis*. Both in *L. Chilensis* and *Platensis*, the sub-orbital foramina are kidney-shaped, the emarginated portion being downwards, whilst in *L. vulgaris* they approach somewhat to a triangular figure, the apex being external. In *L. Chilensis*, however, this foramen is comparatively larger than in *L. Platensis*, and the outer portion of the foramen forms the segment of a larger circle than the inner one, whilst in *L. Platensis* both portions are equal.

The principal difference in the dentition of the La Plata and the Chile otters, consists in the comparatively smaller size of the posterior molars, both of the upper and lower jaws, of the latter species. In the upper jaw, the "carnassière" has its inner lobe, approaching somewhat to a triangular form, whereas in *L. Platensis* it is broader and almost semicircular. In the lower jaw, the last molar but one has the inner lobe much smaller than the middle outer lobe, whilst in *L. Platensis* these two lobes are of nearly equal size and elevation. Other points of dissimilarity will be perceived in the annexed table of admeasurements.

	L. Chilensis. In.	*L. Chilensis.* Lines.	*L. Platensis.* In.	*L. Platensis.* Lines.
Whole length of skull	3	$9\frac{3}{4}$	4	$2\frac{1}{2}$
Greatest width	2	$6\frac{1}{3}$	2	$10\frac{1}{3}$
Width of skull from the apex of one mastoid process to the opposite	2	$3\frac{1}{4}$	2	$8\frac{1}{4}$
Length of palate	1	6	1	10
Breadth of palate between the posterior molars		$7\frac{3}{4}$		$7\frac{3}{4}$
Length from last molar to posterior margin of palate		$3\frac{1}{3}$		$5\frac{1}{2}$
from base of canine to hinder part of last molar		$11\frac{2}{3}$	1	$1\frac{1}{8}$
of carnassière		5		$5\frac{2}{3}$
Width of do.		5		$6\frac{1}{4}$
Length of last molar		$2\frac{3}{4}$		$3\frac{1}{2}$
Width of do.		$4\frac{1}{3}$		$5\frac{2}{3}$
Length of ramus of lower jaw	2	$4\frac{1}{2}$	2	$8\frac{3}{4}$
from canine to hinder portion of last molar (lower jaw)	1	$2\frac{1}{4}$	1	$4\frac{1}{4}$
of last molar but one (lower jaw)		$5\frac{3}{4}$		$6\frac{2}{3}$
Width of do.		$2\frac{2}{3}$		$3\frac{1}{2}$

"These animals are exceedingly common amongst the innumerable channels and bays, which form the Chonos Archipelago. They may generally be seen quietly swimming, with their heads just out of water, amidst the great entangled beds of kelp, which abound on this coast. They burrow in the ground, within the forest, just above the rocky shore, and I was told, that they sometimes roam about the woods. This otter does not, by any means, live exclusively on fish. One was shot whilst running to its hole with a large volute-shell in its mouth; another (I believe the same species) was seen in Tierra del Fuego devouring a cuttle fish. But in the Chonos Archipelago, perhaps the chief food of this animal, as well as of the immense herds of great seals, and flocks of terns and cormorants, is a red coloured crab (belonging to the family Macrouri) of the size of a prawn, which swims near the surface in such dense bodies, that the water appears of a red colour. This specimen weighed nine pounds and a half."—D.

Monoceta Pl. 10

Delphinus Fitz-Roya

FAMILY—DELPHINIDÆ.

DELPHINUS FITZROYI.

PLATE X.

D. suprà niger; capitis corporisque lateribus, corporeque subtùs niveis; caudâ, pedibus, labioque inferiore, nigris; fasciis latis duabus per latus utrumque obliquè excurrentibus, nigréscenti-cinereis, hujusque coloris fasciâ utrinque ab angulo oris ad pedem tendente.

DESCRIPTION.—Upper parts of the body black, under parts pure white, the two blended into each other by gray: extremity of snout, a ring round the eye, the edge of the under lip, and the tail fin, black; dorsal and pectoral fins dark gray; a broad gray mark extends from the angle of the mouth to the pectoral fin; above which, the white runs through the eye and is blended into gray over the eye; two broad deep-gray bands are extended in an oblique manner along each side of the body, running from the back downwards and backwards; iris of eye dark brown. Body anteriorly somewhat depressed, posteriorly compressed; head conical, arched above; the lower lip projecting beyond the upper; eye placed above and behind, but near the angle of the mouth; breathing vent situated in the same line as the eyes—supposing a circle to be taken round the head. Teeth slightly curved, and conical; in the upper jaw twenty-eight in number on each side, and in the lower, twenty-seven.

	Ft.	In.	Lines.
Total length (measuring along the curve of back)	5	4	0
Length from tip of muzzle to vent	3	10	9
to dorsal fin	2	6	5
to pectoral	1	4	5
to eye	0	9	9
to breathing aperture (following curve of head)	0	10	7
to angle of mouth	0	7	9
of dorsal fin along the anterior margin	1	0	5
Height of do.	0	6	4
Length of pectoral, along anterior margin	1	2	8
Width of tail	1	4	5
Girth of body before dorsal fin	3	0	6
before pectoral fin	2	8	2
before tail fin	0	7	8
of head over the eyes	2	0	0

Habitat, coast of Patagonia, Lat. 42° 30′, (*April.*)

E

This species, which I have taken the liberty of naming after Captain Fitz-Roy, the Commander of the Beagle, approaches in some respects to the *Delphinus superciliosus* of the " Voyage de la Coquille," but that animal does not possess the oblique dark-gray bands on the sides of the body ; it likewise wants the gray mark which extends from the angle of the mouth to the pectoral fins. In the figure the under lip of the *D. superciliosus* is represented as almost white, whereas in the present species it is black : judging from the figures, there is likewise considerable difference in the form. The figure which illustrates this description agrees with the dimensions, which were carefully taken by Mr. Darwin immediately after the animal was captured, and hence is correct.

" This porpoise, which was a female, was harpooned from the Beagle in the Bay of St. Joseph, out of several, in a large troop, which were sporting round the ship. I am indebted to Captain FitzRoy for having made an excellent coloured drawing of it, when fresh killed, from which the accompanying lithograph has been taken."— D.

Family — CAMELIDÆ.

Auchenia Llama. *Desmarest.*

Guanaco of the aborigines of Chile.

" The Guanaco abounds over the whole of the temperate parts of South America, from the wooded islands of Tierra del Fuego, through Patagonia, the hilly parts of La Plata, Chile, even to the Cordillera of Peru. I saw several of these animals in Navarin Island, forty miles north of Cape Horn ; the Guanaco, therefore, has, with the exception of a fox and mouse, inhabitants of the same island, the most southern range of all American quadrupeds. Although preferring an elevated site, it yields in this respect to its near relative the Vicuña. On the plains of Southern Patagonia, we saw them in greater numbers than in any other part. Generally they go in small herds, from half a dozen to thirty together ; but on the banks of the Santa Cruz, we saw one herd, which must have contained at least five hundred. On the norther shores of the Strait of Magellan they are also very numerous. The Guanacoes are generally wild and extremely wary : Mr. Stokes told me, that he one day in Patagonia saw through a glass a herd of these beasts, which evidently had been frightened, and were running away at full speed, although their distance was so great that they could not be distinguished by the naked eye.

" The sportsman frequently receives the first intimation of their presence, by hearing from a long distance their peculiar shrill neighing note of alarm. If he then looks attentively, he will, perhaps, see the herd standing in a line on the side of some distant hill. On approaching, a few more squeals are given, and then off they set, at an apparently slow but really quick canter, along some narrow beaten track to a neighbouring hill. If, however, by chance he should abruptly meet a single animal, or several together, they will generally stand motionless, and intently gaze at him ; — then, perhaps, move on a few yards, turn round, and look again. What is the cause of this difference in their shiness ? Do they mistake a man in the distance for their chief enemy the puma ? Or does curiosity overcome their timidity ? That they are curious is certain, for if a person lies on the ground, and plays strange antics, such as throwing up his feet in the air, they will almost always approach by degrees to reconnoitre him. It is an artifice that was repeatedly practised with success by the sportsman of the Beagle, and it had moreover the advantage of allowing several shots to be fired, which were all taken as parts of the performance. On the mountains of Tierra del Fuego, and in other places, I have more than once seen a Guanaco on being approached, not only neigh and squeal, but prance and leap about in the most ridiculous manner, apparently in defiance, as a challenge. These animals are very easily domesticated, and I have seen some in this state near the houses in northern Patagonia, although at large on their native plains. They are, when thus kept, very bold, and readily attack a man, by striking him from behind with both knees. It is asserted, that the motive for these attacks is jealousy on account of their females. The wild Guanacoes, however, have no idea of defence ; and even a single dog will secure one of these large animals, till the huntsman can come up. In many of their habits they are like sheep in a flock. Thus when they see men approaching in several directions on horseback, they soon become bewildered, and know not which way to run. This circumstance greatly facilitates the Indian method of hunting, for they are thus easily driven to a central point, and are encompassed.

" The Guanacoes readily take to the water; several times at Port Valdes they were seen swimming from island to island. Byron, in his voyage, says he saw them drinking salt water. Some of our officers likewise saw a herd apparently drinking the briny fluid from a Salina near Cape Blanco ; and in several parts of the country, if they do not drink salt water, I believe they drink none at all. In the middle of the day, they frequently roll in the dust, in saucer-shaped hollows. The males often fight together; one day two passed quite close to me, squealing and trying to bite each other ; and several were shot with their hides deeply scored. Herds appear sometimes to set out on exploring parties : at Bahia Blanca, where within thirty miles of the coast these animals are extremely scarce, I one day saw the tracks of thirty or forty, which had come in a direct line to a muddy salt water creek. They then must have perceived, that they were approaching the sea, for they had wheeled with the regularity of cavalry, and had returned back in as straight a line, as they had advanced. The Guanacoes have one singular habit, the motive of which is to me quite inexplicable, namely, that on successive days they drop their dung on one defined heap. I saw one of these heaps, which was eight feet in diameter, and necessarily was composed of a large quantity. Frezier remarks on this habit as common to the Guanaco as well as to the Llama ;* he says it is very useful to the Indians, who use the dung for fuel, and are thus saved the trouble of collecting it.

"The Guanacoes appear to have favourite spots for dying in. On the banks of the Santa Cruz, the ground was actually white with bones in certain circumscribed spaces, which generally were bushy and all near the river. On one such spot I counted between ten and twenty heads. I particularly examined the bones ; they did not appear, as some scattered ones which I had seen, gnawed or broken as if dragged together by a beast of prey. The animals in most cases, must have crawled, before dying, beneath and amongst the bushes. Mr. Bynoe informs me, that during the last voyage, he observed the same circumstances on the banks of the Rio Gallegos. I do not at all understand the reason of this; but I may add, that the Guanacoes which were wounded on the plains near the Santa Cruz invariably walked towards the river. This quadruped seems particularly liable to contain in its stomach bezoar stones. The Indians who trade at the Rio Negro, bring great numbers to sell as Remedios or quack medicines ; and I saw one old man with a box quite full of them, large and small."—D.

---

* D'Orbigny says, (vol. ii. p. 69,) that all the species of the genus have this habit.

Family—CERVIDÆ.

Cervus campestris.

Cervus campestris, *F. Cuvier*, in Dict. des Sc. Nat. VII. p. 484.
————————, Cuvier Oss. Foss. IV. p. 51. Pl. 3. f. 46.*
Guazuti, *Azara*, "Natural History of the Quadrupeds of Paraguay." W. P. Hunter's translation,
   vol. i. p. 135.
————————, French translation, vol. i. p. 77.

Besides skins of this species of stag, I find, in Mr. Darwin's collection, three pairs of horns, which, together with a pair belonging to one of the skins, constitute a sufficiently complete series to illustrate the different forms which these appendages assume, as the animal increases in size.

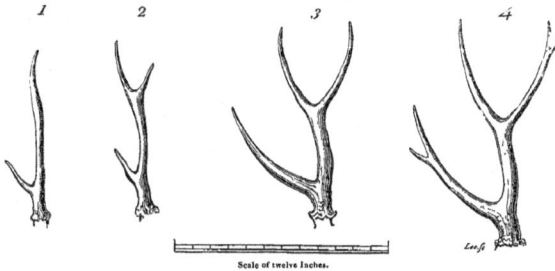

Scale of twelve Inches.

The above four sketches, which are all drawn to the same scale, will help to convey a clear idea of the forms, and relative proportions, of these horns.

The most simple horn (fig. 1.) consists of a *beam*, eight and a half inches long, which is slightly arched outwards and considerably compressed about two and a half inches from the apex. At one inch from the base there is a small brow antler which projects forwards and upwards.

In the next horn, (fig. 2.) there is the same small brow antler, but there is a single small *snag*, about equal in size to the brow antler, which is directed back-

* Figures 47 and 48 of M. Cuvier's work represent horns so unlike either of those brought over by Mr. Darwin, that I cannot help suspecting they belong to some other species of stag.

wards and upwards, and is situated at three and a quarter inches from the apex of the beam. The total length of the beam is eight inches, measured in a straight line.

The third pair of horns, (fig. 3.) which must have belonged to an animal considerably older than either of the preceding pairs, exhibits a large brow antler, in length exceeding half that of the beam : here the posterior snag is also large, and is directed backwards and upwards, whilst the apical portion of the beam is directed forward about as much as the snag is directed backwards. The total length of this horn is eleven and a half inches, measured in a straight line.

The last figure (No. 4.) represents the horn of one of the specimens of which an entire skin was brought over. This horn differs only from the last in being slightly larger, and in having two additional small snags, one springing from the under side, and near the apex, of the brow antler, and the other springing from the hinder part, and near the apex of the great posterior snag.

" The Spaniards say they can distinguish how old a deer is by the number of the branches on the horns. They affirmed that the specimen, of which figure 4 represents one of the horns, was nine years old. It certainly was a very old one, as all its teeth were decayed. This specimen was killed at Maldonado, in the middle of June ; another specimen was killed at Bahia Blanca, (about three hundred and sixty miles southward,) in the month of October, with the hairy skin on the horns : there were others, however, whose horns were free from skin. At this time of the year, many of the does had just kidded. I was informed, by the Spaniards, that this deer sheds its horns every year.

" The *Cervus campestris* is exceedingly abundant throughout the countries bordering the Plata. It is found in Northern Patagonia as far south as the Rio Negro, (Lat. 41°); but, further southward, none were seen by the officers employed in surveying the coast. It appears to prefer a hilly country ; I saw very many small herds, containing from five to seven animals each, near the Sierra Ventana, and among the hills north of Maldonado. If a person, crawling close along the ground, slowly advances towards a herd, the deer frequently approach, out of curiosity, to reconnoitre him. I have by this means killed, from one spot, three out of the same herd. Although thus so tame and inquisitive, yet, when approached on horseback, they are exceedingly wary. In this country nobody goes on foot, and the deer knows man as its enemy, only when he is mounted, and armed with the bolas. At Bahia Blanca, a recent establishment in Northern Patagonia, I was surprised to find how little the deer cared for the noise of a gun : one day, I fired ten times, from within eighty yards, at one animal, and it was much more startled at the ball cutting up the ground, than at the report.

" The most curious fact, with respect to this animal, is the overpoweringly strong and offensive odour which proceeds from the buck. It is quite indescribable: several times, whilst skinning the specimen, which is now mounted at the Zoological Museum, I was almost overcome by nausea. I tied up the skin in a silk pocket-handkerchief, and so carried it home: this handkerchief, after being well washed, I continually used, and it was, of course, as repeatedly washed; yet every time, when first unfolded, for a space of one year and seven months, I distinctly perceived the odour. This appears an astonishing instance of the permanence of some matter, which in its nature, nevertheless, must be most subtile and volatile. Frequently, when passing at the distance of half a mile to leeward of a herd, I have perceived the whole air tainted with the effluvium. I believe the smell from the buck is most powerful at the period when its horns are perfect, or free from the hairy skin. When in this state the meat is, of course, quite uneatable; but the Spaniards assert, that if buried for some time in fresh earth, the taint is removed. These deer generally weigh about sixty or seventy pounds."—D.

FAMILY — MURIDÆ.

1. MUS DECUMANUS.

Mus decumanus, *Auctorum.*

IN the extensive collection of Rodent animals brought home by Mr. Darwin, I find several specimens of the above named species, that is to say, animals which resemble the European specimens of *Mus Decumanus* in all those characters which are the least liable to variation in individuals of the same species, such as the proportions which the various parts of the animal bear to each other: they differ, however, somewhat in colouring.

Buenos Ayres, Maldonado, Valparaiso, East Falkland Island, and Keeling Island, are each, it appears, infested with the common European rat. I have now before me two specimens from East Falkland Island, and one specimen from each of the other localities, and among these I find none equal in size to the largest European specimens: as regards the colouring, the Buenos Ayres specimen differs only from the English specimens of *Mus Decumanus*, in having the upper parts of a richer and deeper hue, owing to the tips of the shorter hairs being of a deep yellow instead of pale yellow, and in having a rusty tint over the haunches.

Mr. Darwin found this variety " common about houses in the country around Buenos Ayres."

In the Maldonado variety, the shorter hairs of the upper parts of the body are of a rusty yellow colour at the apex, in other respects it resembles the British variety. The rusty yellow colour of the tips of the hairs produces a general reddish hue, which is the more conspicuous, when the animal is placed near an English specimen. " Was caught in a house, at Maldonado. I saw a specimen of the common gray English, or Norway rat, lying dead in the streets, and it certainly had a very different appearance from these red rats. The latter, I saw crawling about the hedges in the interior provinces at Santa Fé, and likewise in the forest of the island of Chiloe. This latter fact, however, is a strong argument against its being aboriginal, since I did not find even one undoubted American species, out of the many which I collected, inhabiting both sides of the Cordillera."—D.

The specimen from Valparaiso very closely resembles that from Maldonado; it is, perhaps, a little less red. " Common about the houses in the town of Valparaiso."

The two specimens from East Falkland are of a brighter hue, and have less gray in their colouring, than in the European variety of the common rat. " One of them was caught in a Bay, which is sometimes frequented by shipping, but which is distant thirty or forty miles from any habitation. These rats have spread, not only over the whole of East and West Falkland, but even on some of the outlying islets. When the cold, wet, and gloomy nature of the climate is considered, it is surprising that these animals should be able to find food to live on."—D.

The general hue of the Keeling Island specimen, is deep brown, the longer hairs of the upper parts of the body being, as usual, black ; but the shorter hairs, instead of having the pale yellow tint which we observe in the European, (or, rather, British) specimens of *Mus Decumanus*, are of a deep, rusty yellow. The most remarkable difference, however, consists in the colouring of the under parts being of a yellowish tint, and, towards the root of the tail, of a very distinct buff yellow : the feet are brownish.

" This rat is exceedingly numerous on some of the low coral islets forming the margin of the Lagoon of Keeling Island, in the Indian Ocean. The climate is dry and hot. The rats are known to have come in a vessel from the Mauritius, which was wrecked on one of the islets, which is now called Rat Island. They appeared stunted in their growth, and many of them were mangy. They are supposed to live chiefly on cocoa-nuts, and any animal matter the sea may chance to throw up. They have not any fresh-water ; but the milk of the cocoa-nut would supply its place."—D.

The principal dimensions of the above animals are as follows : —

	Specimen from Buenos Ayres	Maldonado.	Valparaiso.	East Falkland.	East Falkland.	Keeling Island.
	In. Lines.	In. Lines.	In. Lines.	In. Lines.	In. Lines.	In. Lines.
Length from nose to root of tail .	9 9	9 3	8 6	8 9	9 0	8 3
of tail	Imperfect	6 0	6 6	Imperfect	6 0	6 6
of tarsus	1 7	1 7	1 7	1 7	1 7	1 7

Upon comparing the skull of the Valparaiso variety with that of a British specimen of *Mus decumanus*, I could perceive no difference. A skull from West Falkland did not differ, neither did the dentition of the Keeling Island specimen above noticed. A perfect specimen of this last I have not had an opportunity of examining.

2. MUS (DECUMANUS *var.* ?) MAURUS.

Mus maurus, *Waterh.* in Proceedings of the Zoological Society of London, for February, 1837, p. 20.

M. pilis suprà purpurescenti-nigris; subtùs plumbeis; auribus parvulis, pallidè fuscis: caudâ corpus ferè æquante.

DESCRIPTION. — The character of the fur of this animal nearly resembles that of *Mus decumanus;* it is, however, of a harsher nature: the general colour of the upper parts and sides of the body is purple-black, arising from the longest hairs being of this colour, and likewise the tips of those which are next in length; the latter, however, excepting at the tip, are white, and this white is not entirely hidden, even when the hairs are in their ordinary position : on the head the hairs assume a brownish hue, and are tolerably uniform : the limbs, and under parts of the body, are of a deep gray colour, with a faint purple-brown wash: the under fur is gray : the ears are small, of a brown-white, or very pale brown colour, and furnished with minute brown hairs : the small, scattered, bristly hairs of the tail are of an uniform brownish-black colour. The hairs of the moustaches are black at the base, and grayish at the apex.

	In. Lines.			In. Lines.
Length from nose to root of tail . . 11 3		Length of ear 0 0¼		
of tail 7 6		from nose to ear 2 2		
of tarsus 1 8				

Habitat, Maldonado, La Plata, (*June.*)

This rat is very closely allied to *Mus decumanus*, and I think may possibly prove an extraordinary local variety of that animal. Having but one skin, and no skull, I am unable to satisfy myself on this point. Its size, as will be seen by the admeasurements, exceeds that of the common rat, or, rather, it exceeds ordinary specimens of that animal, for I have seen *some* which were equal to it.

"It was killed near Maldonado, where it frequented holes in the sand hillocks near the shore. It is likewise found on the island of Guritti. If ships are ever infested with these monstrous rats, the above-mentioned localities are very likely places to have received colonies by such means. An old male weighed fifteen ounces and three quarters. The ears of this rat, when alive, were of a pale colour, which made a singular contrast with the black fur of its body."—D.

3. Mus Jacobiæ.

Mus decumanoïdes,* *Waterh.* in " Catalogue of the Mammalia preserved in the Museum of the Zoological Society of London."

M. suprà fuscus, griseo-lavatus, subtùs albus: pedum pilis sordidè albis; caudâ corpore cum capite paulò longiore; auribus mediocribus: pilis perlongis in dorso crebrè inter cæteros commixtis.

DESCRIPTION. — The general tint of the upper parts of this rat, is grayish-brown, (very nearly resembling that of *Mus decumanus*); the longest hairs, which on the hinder portion of the back are one inch and a half in length, are black; the ordinary hairs are black at the apex, there is then, on each hair, a considerable space occupied by pale yellow, and the remaining, or basal portion, is grayish white; the under fur is gray: the hairs of the chin, throat, and under parts of the body, are white, and without any gray colour at the roots: the feet are covered with dirty grayish hairs: the tail, which is slender, is very sparingly furnished with minute black hairs, both above and beneath: the ears are of moderate size, of a brownish flesh colour, and, to the naked eye, appear to be destitute of hair. The hairs of the moustaches are most of them black at the base, and grayish at the apex.

* The MS. name of *M. decumanoïdes,* which I had applied to this animal, has been changed, in consequence of my having seen a different species, with the same name attached, in the museum of the India House.

	In. Lines.			In. Lines.
Length from nose to root of tail	. . 7 6	Length of ear	. ˙ . . . 0 7½
of tail 7 6		from nose to ear 1 7½
of tarsus 1 4¼		

Habitat, James Island, Galapagos Archipelago, Pacific Ocean, (*October.*)

This species is scarcely equal in size to a full grown common black rat, (*Mus Rattus*), the head is rather shorter in proportion, the tarsi are smaller, and the tail is longer. In the character of the fur, and length of the hairs, it *very* closely resembles that species : the ears are larger than in *M. decumanus*, and about equal to those of *M. Rattus*. In having the hairs of the under parts of the body of an uniform colour, (i. e. not gray at the base,) it resembles the *Mus Tectorum* of Savi ; but the large size of that animal, the greater length of the fur, and its colouring, all serve to distinguish it from the present species, which I may here observe, is truly an old world form, and very distinct from another species, also from the Galapagos, which is hereafter described.

" It is very common in James Island, but is not found on all the islands, if on any other in the Archipelago. Although its appearance is so like that of the common rat, yet its habits appear to be rather different : it is less carnivorous, and does not appear to be so strongly attached to the habitations of man. This island was frequented, about one hundred and fifty years since, by the vessels belonging to the Bucaniers ; so that the common rat might easily have been transported here. And if a very peculiar climate, a volcanic soil, and strange food, can together produce a race, or strongly marked variety, there is every probability of such change having taken place in this case."—D.

4. Mus (Rattus var.?) insularis.

M. suprà grisescenti, colore subtùs dilutiore; tarsis purpureo-nigris: caudâ corpus cum capite æquante: auribus mediocribus: vellere molli.

Description.—No. 1. The general colour of this animal is what might be termed black, there is, however, an obscure purple-brown hue on the upper parts of the body, and the sides and under parts have a grayish tint, the hairs covering the feet above are of an uniform deep purple-brown, almost black. All the hairs of the body are gray at the base : the hairs of the moustaches are long and numerous, and of a black colour, having one or two white hairs intermixed : the ears are of moderate size, and very sparingly furnished with minute dark hairs: the tail is long and slender, and has small, scattered, bristly hairs, of a brown-black colour.

	In. Lines.			In. Lines.
Length from nose to root of tail	7 0	Length of ear	. . . .	0 7
of tail	6 6	from nose to ear	. . .	1 6
of tarsus	1 3½			

No. 2. Hairs along the centre of the back chiefly black, and but obscurely annulated, near the apex, with deep yellow: towards the sides of the body, and over the haunches, the hairs are more distinctly annulated, and on the sides of the body they are of a pale yellow at the apex: on the under parts the hairs are gray, tipped with dirty yellowish white: the feet are of the same deep purple-brown hue as in the specimen first described.

Habitat, Ascension Island, Atlantic Ocean, (*July.*)

These two animals not only differ in the colour of the fur, one being of a grizzled brownish colour, and the other black, but there is a considerable difference in the texture of the fur. In the black specimen, the fur is very soft and glossy, and the long hairs, which are abundant, are very slender. In the brown specimen, the fur is of a harsher nature, the long hairs are not so abundant, but longer, and less slender. On the other hand, they agree in size, dentition, the length of the head, tarsus, and ears, and differ but in a trifling degree (about three lines,) in the length of the tail.

Upon comparing the Ascension Island specimens with *M. Rattus*, I find that, although in size they are about one third less, yet the teeth precisely agree, not only in form, but in size. The relative proportions of the head, ears, and tarsi, also agree. Besides the general colouring of the fur, they both differ in having the hairs of the feet uniformly purple-black, those in *Mus Rattus* being much paler, and even whitish, on the toes. In the character of the fur, there is much difference. The long silky hairs, which are so conspicuous in *Mus Rattus*, are replaced, in the black specimen, by hairs which are scarcely to be distinguished from the ordinary fur; and in the other specimen, although rather longer and more distinct, they are short, compared with those of the black rat.

" The specimen which has a black, and glossy fur, frequents the short coarse grass near the summit of the island, where the common mouse likewise occurs. It is often seen running about by day, and was found in numbers, when the island was first colonized by the English, a few years since. The other, and browner coloured variety, lives in the out-houses near the sea-beach, and feeds chiefly on the offal of the turtles, slaughtered for the daily food of the inhabitants. If the settlement were destroyed, I feel no doubt that this latter variety would be compelled to migrate from the coast. Did it originally descend from the summit? and, in the case just supposed, would it retreat there? and, if so, would its black colour return? It must, however, be observed, that the two localities are separated from each other by a space, some miles in width, of bare lava and ashes. Does the summit of Ascension, an island so immensely remote from any continent, and the summit itself surrounded by a broad fringe of desert volcanic soil, possess a small quadruped, peculiar to itself? Or, more probably, has this new species been brought, by some ship, from some unknown quarter of the world? Or, I am again tempted to ask, as I did in the case of the Galapagos rat, has the common English species been changed, by its new habitation, into a strongly marked variety?"—D.

Mr. Darwin seems to have foreseen the difficult problem which these two rats have furnished, and although I have spent much time in studying the Muridæ, I must confess I have been exceedingly puzzled by the animals in question. It appears as if the brown, and black rats, (*M. decumanus*, and *M. Rattus*,) and likewise the common mouse, (*M. Musculus*,)* all of which follow man in his peregrinations, and which, to a certain degree, are dependent upon man, and may therefore be termed semi-domestic animals; like *really* domestic animals, are subject to a greater degree of variation than those species which hold themselves aloof from him.

Upon the whole then I have determined to describe the two Ascension Island specimens as one species, and as varieties of the *Mus Rattus*, but with a mark of doubt, since I do not possess sufficient materials for a rigorous examination, having, in fact, but one skin of each variety, and neither skull nor skeleton. I have also applied the name of *insularis*, to designate this variety or species, whichever it may be, for, supposing it be not a distinct species, it is so marked a variety, that a name for it is desirable.

---

* The great Bandicoot rat of India, (*Mus giganteus*, of Hardwicke,) ought, perhaps, to be added to the species above enumerated; and I strongly suspect several catalogued species will prove but varieties of this animal.

5. Mus Musculus.

Mus Musculus, *Auctorum.*

Of this species, there are six specimens in Mr. Darwin's collection; two were found " living in the short grass, near the summit of the Island of Ascension, where the climate is temperate."—D. Two others were procured " on a small, stony, and arid island, near Porto Praya, the capital of St. Jago, in the Cape de Verde Islands,—climate very hot and dry. Excepting during the rainy season, which is of short duration, these little animals can never taste fresh water, nor does the island afford any succulent plant."—D. A specimen was also procured " on a grassy cliff, on East Falkland Island, at the distance of a mile from any habitation. It is singular that so delicate an animal should be able to subsist under the cold, and extremely humid climate, of the Falkland Islands, and on its unproductive soil."—D. These specimens are all of them rather less than full grown individuals of the same species procured in England; in other respects, they do not differ.

The sixth specimen, which is from Maldonado, is considerably less than British specimens of the common mouse, and is of a richer and brighter colour, the head is smaller, the muzzle shorter in proportion, whilst the tarsi are even longer than in a large specimen of *M. Musculus.* These points of dissimilarity induced me to believe it was a distinct species, and to apply to it the specific name of *brevirostris.** Upon re-examination, with the advantage of more experience, and consequently a better knowledge of the characters of these animals, I have changed my opinion. The teeth indicate that it is not an adult specimen, and agree perfectly with 'those of *M. Musculus*, both in form and size. " Common in the houses of the town of Maldonado, and its habits are similar to those of *Mus Musculus.*" — D.

* See Proceedings of the Zoological Society for February 14th, 1837, p. 19.

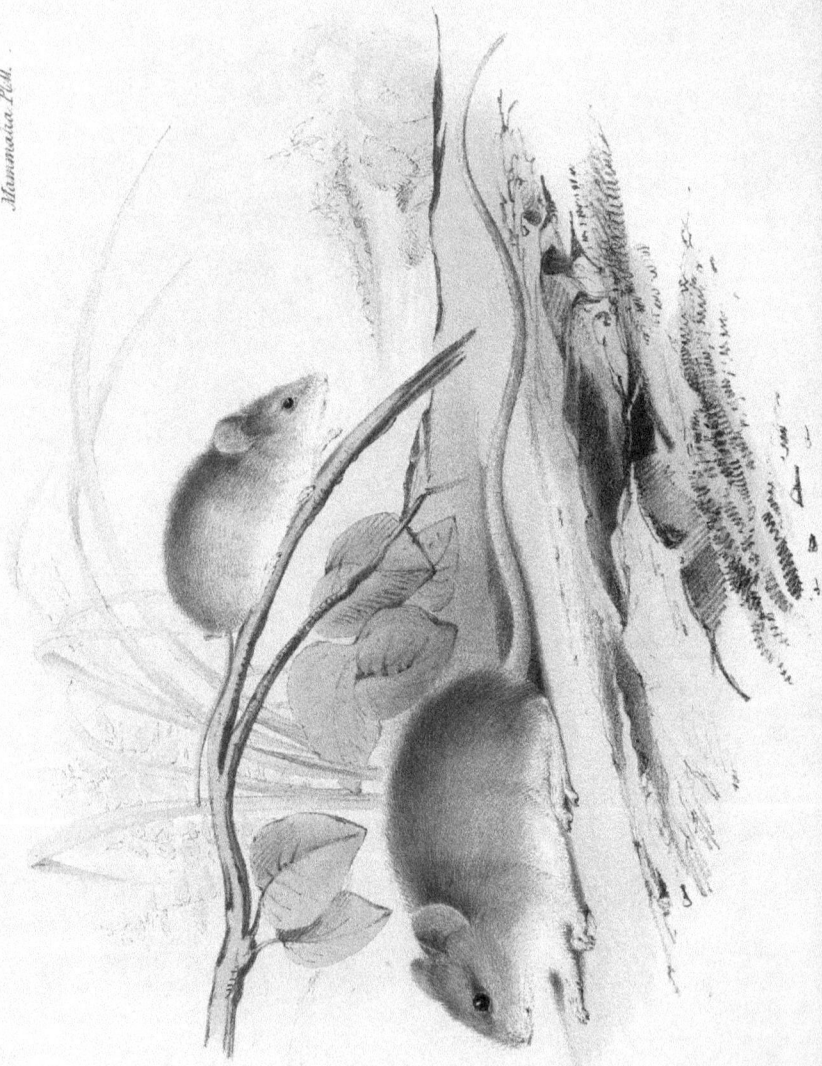

6. MUS LONGICAUDATUS.

PLATE XI.

Mus longicaudatus, *Bennett*, Proceedings of the Committee of Science and Correspondence of the Zoological Society of London for January, 1832, p. 2.

M. pallidè flavescenti-fuscus ; corpore subtùs albo, levitèr flavo lavato ; pedibus albis ; tarsis permagnis ; caudá perlongá ; auribus parvulis.

DESCRIPTION.—Fur long and soft; general colour pale yellow-brown, the hairs of the ordinary fur being fulvous near the apex, and the longer hairs brown. On the sides of the body, cheeks, and external side of limbs, the fulvous hue prevails. The inner side of the limbs and the under parts of the body are white, but have an indistinct yellowish hue. All the hairs of the body are of a deep gray colour at the base. The ears are small, well clothed with hairs ; those on the inner side are chiefly yellow ; externally, on the fore part they are brown, and posteriorly whitish. The feet are of a flesh-colour, and furnished above with white hairs ; the tarsi are but sparingly provided with minute hairs on the upper side, and are naked beneath : they are of unusually large size. The fore feet are of moderate* size, and furnished with a very large carpal tubercle. The tail is very nearly double the length of the body, if the latter be measured in a straight line ; it is of a brownish flesh-colour above, paler beneath, and sparingly furnished with minute bristly hairs ; those on the upper surface being brown, and on the under side white. The hairs of the moustaches are long, of a black colour, and grayish at the apex.

	In.	Lines.		In.	Lines.
Length from nose to root of tail	3	9	Length of tarsus (claws included)	1	1
of tail	5	3	of ear	0	4
from nose to ear	0	10½			

Habitat, Chile.

* As I shall have occasion to use the terms *moderate, long, short, large,* &c. it may be well to state that I take the common mouse, (*Mus Musculus,*) as my standard of comparison. The ears, feet, tail, length of the fur, general proportions, &c. are in that animal what I term moderate.

The most conspicuous characters of the present species consist in the immense length of the tail, and the great size of the hinder feet. * It is about equal in size to *Mus Musculus;* its form, however, is somewhat stouter; in colour it is much paler and brighter.   The head is larger in proportion ; the ears are smaller, and more densely clothed with hair; the fore feet are rather larger, and the fleshy tubercle on the under side of the wrist is also larger.   The thumb nail is flattened, and rounded at the tip, as in *Mus Musculus,* but is longer, and more distinct than in that animal.

The skull of *M. longicaudatus,* (Plate 34, Fig. 1,) is considerably larger than that of the common mouse, but in form scarcely differs from it; its upper surface is rather more convex, and the interparietal bone proportionately less.   The length of the skull is 1 inch ; breadth, $6\frac{1}{2}$ lines ; distance between the fore part of the incisor, and the first molar of the upper jaw, $3\frac{1}{2}$ lines.   The dentition is figured in Plate 34, Figs. 1. *b* and 1. *c.*

The above account is drawn up from the same specimen as that from which Mr. Bennett took his description, and which was brought from Chile by Mr. Cuming, who states that the animal in question lives in trees, and constructs its nest with grass.

In Mr. Darwin's collection, I find an animal which agrees in all the more important characters with the one above described, but differs in being of a deeper colour, (approaching more nearly, in this respect, to the common mouse,) and in having the tail a trifle shorter.   The skull is about $\frac{3}{4}$ of a line shorter, but its proportions agree precisely : the proportions of the feet, and the general form of the animal, also agree.   This specimen is likewise from Chile, (Lat 37° 40',) and, according to Mr. Darwin, " overran the wooded country south of Concepcion, in swarms of infinite numbers.   Captain FitzRoy, on his return from visiting the wreck of H. M. S. Challenger, had the kindness to bring me this specimen.   So destructive was this little animal, that it even gnawed through the paper of the cartridges belonging to the people who were wrecked."—D.

---

* A long tarsus is generally accompanied by a proportionately long tail. I presume that those Mice which have long tarsi are in the habit of making great leaps, and that in these leaps, the tail serves to steady and balance the body.

Mammalia. Pl. 12.

Mus elegans

Mus bimaculatus

Mus elegans.

Plate XII.

Mus elegans, *Waterh.*, Proceedings of the Zoological Society of London for February 1837, p. 19.
Eligmodontia typus, *F. Cuvier*, Annales des Sciences Naturelles for March 1837. Tom. 7. p. 169. Pl. 5.

M. suprà flavus, vellere pilis fuscescentibus adsperso, his ad latera, et prope oculos rarioribus; pilis pone aurem utramque, labiis, corpore subtùs, pedibusque niveis; auribus magnis; caudâ capite corporeque paulo longiore; tarsis longis subtùs pilis obsitis.

DESCRIPTION.—Fur very long and soft ; general colour of the upper parts of the body pale brownish yellow ; the lower portion of the cheeks, and the under parts of the body pure white : the hairs of the ordinary fur of the back are gray at the base, pale ochre near the apex, and brown at the apex ; the longer hairs are brownish. On the sides of the body where the longer hairs are less numerous, the pale ochre colour prevails ; the hairs on this part as on the back are deep gray at the base, but at a short distance from the apex they are white ; nearer the tip shaded into yellow, and at the tip brownish : the limbs externally are of a pale yellow colour. The hairs of the throat and chest are pure white to the root, those on the belly are obscurely tinted with gray at the root. The feet are of a pale flesh-colour, and furnished with white hairs ; the fore feet are of moderate size ; the thumb nail is small and rounded, and the carpal tubercle is covered with hairs ; the tarsi are long, and the white hairs extend over the whole of the under parts ; the under side of the toes, however, are but sparingly furnished. There appears to be but one large tubercle on the under side of the tarsus, and this, which is situated near the base of the toes, is thickly covered with silvery-white hairs. The tail is long, pale brown above, and pale flesh-colour beneath ; above, it is furnished with minute brown hairs, and on the under side with white hairs. The ears are rather large, of a pale flesh colour, tolerably well clothed with hairs, which are of a pale yellow colour on the inner side, and white on the outer side — excepting on the fore part, where they are brown. A small tuft of white hairs springs from the base of the ear posteriorly. The hairs of the moustaches are moderate ; black at the base, and grayish at the apex.

G

	In. Lines.				In. Lines.
Length from nose to root of tail .	. 3 7		Length of tarsus	. 0 10
of tail	. 3 9		of ear	. 0 6
from nose to ear . .	. 1 0			

Habitat, Bahia Blanca, (*September.*)

Upon comparing the skull (Pl. 34, fig. 2, *a*.) of *M. elegans* with that of *M. Musculus*, the most evident points of distinction consist in the greater proportionate length of the nasal and frontal bones, and the slenderness of the zygomatic arch in the former animal. Length of skull 11 lines, width 6 lines, distance between front molar and outer side of incisors of upper jaw 3⅝ lines, length of nasal bones 4⅜ lines.

The dentition is figured in Pl. 34, figs. 2. *b*, and 2. *c*.

" Whilst bivouacking one night on shore, amongst some sand hillocks, this mouse, with its tail singed, leapt out of a bush which was placed on the fire. Its hind legs appeared long in proportion to the front, and it did not appear to be very active in endeavouring to make its escape."—D.

Mus elegans is about equal in size to *M. Musculus;* the head is larger in proportion than in the latter, the ears are slightly larger, the tail is longer, and so are the tarsi. The large ears, long tail, and comparatively large size of the feet, combined with the greater size of the animal itself, will render it easy to distinguish this species from *M. gracilipes* and *M. bimaculatus*. From the last mentioned animal it moreover differs in having the head larger in proportion, the fur longer, and the colouring of the upper parts of the body somewhat darker. The white fur is almost confined to the under parts of the body, and there is but a small tuft of white hairs behind the ears, whereas in *M. bimaculatus*, the white fur extends considerably on the sides of the body, the outer side of the limbs are white, and there is a large and conspicuous white spot behind each ear.

In *M. elegans* the whole sole of the tarsus and the carpal tubercles are covered with hair. In *Mus bimaculatus* the hinder *half* of the tarsus only is covered with hair, and in *M. gracilipes* both the hinder half is covered, and there are some scattered hairs extending almost to the two tubercles, which are situated at the base of the longer toes.

The genus *Eligmodontia* of M. F. Cuvier, founded upon a species of mouse from Buenos Ayres, possesses nearly the same characters as the subgenus *Calomys*, established by me in the Proceedings of the Zoological Society for February 1837, and which included the animal above described, and two other species (*M. bimaculatus* and *M. gracilipes*). M. Cuvier's genus is distinguished by there being only one large tubercle on the under side of the tarsus, and in having the carpal pad covered with hair as well as the pad of the tarsus. In these characters our present animal agrees, as it does also in size and in the relative proportions of the tail and tarsus, circumstances which induce me to believe they are identical.

In *M. bimaculatus* and *M. gracilipes* there are six naked tubercles on the under side of the tarsus, and the carpal pad is also naked. In having, however, the tarsus hairy beneath,* in dentition and in colouring, they agree so closely with *M. elegans* that I think they cannot be separated generically.

Mus bimaculatus.

Plate XII.

Mus bimaculatus, *Waterh.*, Proceedings of the Zoological Society of London for February 1837, p. 18.

M. vellere pallidè ochraceo, pilis nigricantibus adsperso, his ad latera rarioribus; rostri lateribus, notâ magnâ pone aurem utramque, artubus, corporeque subtùs niveis; auribus mediocribus; caudâ, quoad longitudinem, corpus fere æquante; tarsis ad calcem pilis argenteo-candidis obsitis.

Description.—Upper parts of the body of a very pale ochre colour, the longer hairs, however, are black, and at the apex grayish, and where they are numerous, as on the back and upper surface of the head, they give greater depth to the colouring; the cheeks and sides of the body are of an almost uniform pale, but bright yellow; the sides of the muzzle, the lower half of the cheeks, the lower portion also of the sides of the body, and the whole of the under parts, are pure white—each hair being uniform in colour to the root, and not, as is usually the case, *gray* at the root. There is likewise a large patch of pure white hairs behind each ear. The feet and tail are of a pale flesh-colour, and furnished with white hairs, with the exception of those on the upper surface of the latter, which are pale brown. The ears are also pale flesh-colour, clothed internally with yellow hairs; externally on the fore part, the hairs are brownish, and on the hinder part, white—they are rather large, and so are the feet. The tail is about equal to the body in length. The hairs of the moustaches are numerous and slender, and most of them are black at the base, and gray at the apex. The hinder half of the tarsus

* In *Mus leucopus* of North America the tarsus is hairy beneath, and in the character of the teeth this animal also agrees with the species above mentioned.

G 2

beneath is covered with minute silvery-white hairs; beside the ordinary tubercles, the anterior portion of the sole of the foot and the base of the toes beneath, are crowded with small rounded warts, which are much more numerous and conspicuous than in the common mouse.

	In. Lines.		In. Lines.
Length from nose to root of tail . .	3 1	Length from nose to base of ear . .	0 8¾
of tail	1 11	of tarsus (claws included) . .	0 8
from nose to eye . . .	0 4½	of ear	0 4½

Habitat, Maldonado, La Plata, (*June.*)

The skull of this animal, is rather shorter and broader than that of *Mus Musculus*, the upper surface is more arched, the zygomatic arch is much more slender, and the nasal bones are rather broader. In the convexity of the upper surface, and the slenderness of the zygomatic arch, this skull very nearly resembles that of *M. gracilipes*; this latter, however, has the zygomatic arch more convex, projecting more suddenly on the anterior part, and the interparietal bone smaller. Length of skull 10 lines, width 5½, length of nasal bones 4 lines, distance between the outer side of the incisors, of the upper jaw, and the first molar 2⅞ lines. See Plate 34, fig. 3. *a*.

The dentition is figured in Plate 34, figs. 3. *b* and *c*.

This mouse is rather less than *M. Musculus*, the tail is much shorter in proportion, the fur is longer and softer, and the ears are more distinctly clothed with hair.

The pale and delicate yellow colour of the upper parts of the body, and the pure white of the under parts, renders the present species conspicuous amongst its congeners. I may further remark that the white colour which in the Muridæ (when it occurs) is usually confined to the under part of the body, or extends but slightly on the sides, is in the present animal extended considerably on the sides of the body, and occupies an equal portion with the yellow of the upper parts. The name *bimaculatus* is applied to this animal on account of the two conspicuous white patches, which are situated behind the ears.

In affinity as well as in appearance it most nearly approaches to *Mus gracilipes* and *M. elegans*; with no other species of the genus Mus, here described, can it be confounded, since these only have the tarsus hairy beneath.

The principal points of distinction between the present animal and *Mus elegans*, are noticed in the account of that species.

"This mouse, when alive, had a very elegant appearance. A countryman, who brought it me, found six of them living together in one burrow."—D.

Mus gracilipes.

Plate XI.

Mus gracilipes, *Waterh.*, Proceedings of the Zoological Society of London, for February 1837, p. 19.

M. suprà flavo-lavatus; *pilis pone aurem utramque, labiis, corporeque subtùs, albis; pedibus parvulis, gracilibus, carneis, suprà et ad calcem pilis albis tectis; caudâ gracili, pilis albis instructâ, quoad longitudinem corpus ferè æquante; auribus mediocribus; vellere mediocri et molli, pilis omnibus ad basin plumbeis.*

Description.—General colour very pale yellowish brown, a tint produced by the admixture of black and pale fawn colour; the hairs of the ordinary fur being of the latter tint near the apex, and dusky at the apex, whilst the longer hairs are black. The feet, tail, under parts of the body and the sides of the muzzle, are pure white. All the hairs of the body, (which are soft, and of moderate length), are deep gray at the base. The ears are of moderate size, well clothed with hairs, of which those on the inner side are yellowish, and those on the outer, are brown on the anterior part, and white on the posterior. A small tuft of white hairs springs from the neck immediately behind the ears; this tuft is hidden when the ears are folded back. The tail is slender and short, (being not quite equal to the body in length) of a pale flesh-colour, and sparingly furnished with minute white hairs. The feet are very small and slender, and the naked parts are of a pale flesh-colour. The sole of the foot is covered with hairs; the toes beneath, and the tubercles (which are as in *Mus Musculus*), however, are naked. The hairs of the moustaches are of moderate length, and of a blackish colour, some of them, however, are grayish white.

	In.	Lines.			In.	Lines.
Length from nose to root of tail .	2	10	Length from nose to ear . .	0	8¼
of tail	1	7	of tarsus (claws included) .	0	6½
from nose to eye . .	0	4½	of ear	0	4¼

Habitat, Bahia Blanca, (*September.*)

This species slightly exceeds the harvest mouse (*Mus messorius*) in size, its ears are considerably larger in proportion, and the tail is shorter. Compared with the common mouse (*Mus Musculus*) it is smaller, the tail is more slender, and shorter, and the feet are likewise more slender and proportionately much smaller; the ears are more distinctly clothed with hairs.

The principal points of distinction between this and the two preceding species are pointed out in the account of *M. elegans.*

Upon comparing the skull of *M. gracilipes* (Pl. 34, fig. 4. *a.*) with that of *Mus Musculus,* the most striking differences consist in its shorter and broader form, the upper surface being more arched, the interparietal bone has a relatively smaller antero-posterior diameter, the occipital region is more convex, and continued more gently and gradually into the upper region of the skull. The zygomatic arch, which is unusually slender, is more dilated (especially on the anterior part) thus giving a squareness to the general form. The nasal bones are not so much attenuated posteriorly. The length of the skull is $8\frac{7}{8}$ lines, the greatest width is $5\frac{1}{8}$ lines, and the distance between the outer side of the incisors and the front molar is $2\frac{3}{4}$ lines.

The dentition is figured in Plate 34, figs. 4. *b* and 4. *c.*

"This specimen was given me by Mr. Bynoe, the surgeon of the Beagle, who caught it amongst some long dry grass."—D.

Mus flavescens.

Plate XIII

Mus flavescens, *Waterh.*, Proceedings of the Zoological Society of London, for February 1837, p. 19.

M. suprà colore cinnamomeo, lateribus capitis, corporisque, æquè ac pectore, auratis; gulâ abdomineque flavescenti-albis: pedibus sordidè albis: auribus mediocribus rotundatis, pilis flavis obsitis: caudâ, corpore, capiteque longiore, suprà fuscâ, subtùs sordidè albâ: tarsis longis.

DESCRIPTION.—Fur long and moderately soft; general colour of the upper parts bright brownish yellow; on the sides of the head and body bright yellow; towards the rump of a deeper hue, and inclining to orange; under parts pale yellow, or yellow-white; chest yellow. The fur both of the upper and under parts of the body deep plumbeous at the base. Feet flesh colour, covered above with white hairs: tarsi long, naked beneath. Ears small, tolerably well clothed with hairs; those on the inner side yellow, but many of them blackish at the base; on the outer side, the hairs are blackish on the fore part and yellow on the hinder part. The hairs of the ordinary fur of the back are of a deep rich yellow colour at the tip, and the longer hairs are blackish. The tail is long, deep brown above and whitish beneath; the hairs of the

Mus Fawnecus

Mus xanthocta

Mus. Magellanicus.

Mus. Brachiotis.

moustaches are rather short and slender, and of a brownish colour. Thumb nail small and rounded.

	In.	Lines.							In.	Lines.
Length from nose to root of tail	.	.	3	9	Length of tarsus	.	. .	.	1	0½
. of tail	.	4	1½		of ear	.	. ·.	.	0	4½
from nose to ear	.	.	1	0						

Habitat, Maldonado, La Plata, (*June.*)

This species is slightly larger than the common mouse; the head is rather larger in proportion; the ears are rather smaller and more distinctly clothed with hair; the tail and tarsi are much longer in proportion. Its bright yellow colouring and proportions distinguish it from any of the species described in this work. Of this animal I do not possess the skull, nor of the teeth do I possess more than the first and second molars of the upper jaw, and the second and last of the lower jaw. These are figured in Plate 34, figs. 5. *a*, and 5. *b*.

Mus Magellanicus.

Plate XIV.

Mus Magellanicus, *Bennett*, Proceedings of the Zoological Society of London for December 1835, p. 191.

M. suprà fuscus, subtùs cinerescenti-albus, pallidè flavo lavatus; auribus mediocribus pilis fuscis obsitis; caudâ corpus caputque æquante; tarsis longis, pilis sordidè albis obsitis.

Description.—Fur very long and moderately soft, general colour deep brown; the hairs of the ordinary fur are gray, tipped with yellowish brown; the longer hairs are black; the sides of the body are yellowish; the under parts are gray-white with a faint yellowish tint, each hair being gray tipped with yellowish white. The ears are rather small, well clothed with hairs; those on the inner side are blackish tipped with yellow, and on the outer side they are blackish on the fore part and dusky on the hinder part. The fore feet are of moderate size, the thumb nail is short and rounded; the tarsi are rather long; both fore and hinder feet are of a brownish colour, and covered above with dirty gray hairs. The tail rather exceeds the head and body in length; it is brown above and dirty white beneath. The hairs of the moustaches are numerous and long, of a brownish colour at the apex and black at the base.

	In.	Lines.						In.	Lines.
Length from nose to root of tail .	.	4	3	Length of tarsus	.	.	.	1	1
of tail	.	4	2	of ear .	.	.	.	0	5
from nose to ear . . .	1	0¼							

Habitat, Port Famine, Strait of Magellan.

This mouse is larger than *Mus Musculus;* the tail is rather longer in proportion; the tarsi much longer; the ears are not quite so large in proportion to the
head, (which greatly exceeds that of *Mus Musculus* in size,) and they are densely
clothed with hair. The fur is longer. In colour, the animal here described is
rather darker than the common mouse. I have one specimen however before me
which *very nearly* agrees in this respect.

The dentition is figured in Plate 34, figs. 6, *a.* and 6, *b.*

From the attention which Mr. Darwin bestowed upon the Muridæ of the
southern parts of South America, I presume his collection affords materials for a
tolerably complete monograph of the species of that portion of the globe. The
species above described, however, does not occur in Mr. Darwin's collection, but
is here introduced in order to make the work more complete, and that I might
more clearly point out the distinctions which exist between it and other
species here described, the account given by Mr. Bennett in the Proceedings
being very short.

Mus arenicola.

Plate XIII.

Mus arenicola, *Waterh.*, Proceedings of the Zoological Society of London, for February 1837, p. 18.

*M. suprà fuscus, subtùs cinerescenti-albus, pallidè flavo tinctus; auribus mediocribus
rotundatis, pilis flavis fuscisque obsitis; caudâ quoad longitudinem corpus æquante;
pedibus cinerescenti-albis: tarsis mediocribus.*

Description.—Fur long, moderately soft; general colour deep brown; sides of
the body with a very obscure yellowish hue; under parts dirty gray with a
faint yellow tint. All the fur deep gray at the base; the hairs of the upper
part of the body obscurely annulated with yellowish brown near the apex,
and dusky at the apex; the longer hairs are black. Feet brownish, covered
above with brown-white hairs; tarsi short. Tail short, blackish above,
brown-white beneath. Ears small, well clothed with hairs; those on the

inner side are yellow at the apex and gray at the base; on the outer side they are of a brownish colour, and on the fore part blackish. The hairs of the moustaches are short and slender, and of a brownish colour. The head is large.

	In.	Lines.			In.	Lines.
Length from nose to root of tail	4	3	Length of tarsus (claws included)	.	0	10
of tail	2	9	of ear	. . . .	0	4½
from nose to ear	1	0				

Habitat, Maldonado, La Plata, (*June.*)

This species is rather larger than the common mouse; its head is proportionately larger, the ears are smaller, the tail considerably shorter, and the fur longer, and in colouring it is a little darker. In size and colour it resembles *M. Magellanicus,* but the shorter tail and tarsi, and the smaller size of the ears will serve to distinguish it.

The skull of *Mus arenicola,* Plate 34. fig. 7, *a,* is rather larger than that of *Mus Musculus,* the nasal portion is broader, the interparietal bone is much smaller, especially in antero-posterior extent; the zygomatic arches are more slender, and the incisive foramina are broader. The horizontal ramus of the lower jaw (Pl. 34. fig. 7, *d.*) is rather less curved, the coronoid process is more elongated, and the condyloid is narrower and also larger. The length of the skull is 11 lines and a half; the width is 6½ lines. The molars of the upper jaw are figured in plate 34 fig. 7, *b.* and those of the under jaw, fig. 7, *c.*

" This specimen was caught on the open grassy plain, by a trap baited with a piece of bird; it is, however, very abundant in the sand hillocks near the coast of the Plata."—D.

13. Mus brachiotis.

Plate XIV.

Mus brachiotis, *Waterh.,* Proceedings of the Zoological Society of London for February 1837, p. 17.

M. suprà obscurè fuscus, subtùs obscurè griseo tinctus; pedibus griseo-fuscis; auribus parvulis; caudâ quoad longitudinem, corpus ferè æquante: vellere longo et molli.

Description.—Fur soft, very long, and dense; ears very small; general colour brown: the hairs of the upper parts, and sides of the head and body are of

H

a deep gray at the base, black at the apex, and narrowly annulated with deep yellow near the apex ; on the throat and belly they are of a paler gray at the base, and grayish white at the apex. The ears are well clothed with brown hairs both within and without, and are for the most part hidden by the long fur of the head. The hairs covering the upper side of the feet are of a palish ashy-brown colour, and the fleshy portion appears to have been brown. The tail is well clothed with hairs, so that the scales are scarcely visible ; on the upper side of the tail the hairs are brownish-black, and on the under side, they are dirty white. The incisors are very slender ; those of the upper jaw are of a very pale yellow colour, and those of the lower are white, or nearly so. The muzzle is slender, and pointed.

	In.	Lines.		In.	Lines.
Length from nose to root of tail . .	4	9	Length of tarsus (claws included) . .	0	11
of tail	2	8	of ear	0	3
from nose to base of ears . .	1	2			

Habitat, Chonos Archipelago, (*December.*)

This mouse is considerably larger than *Mus Musculus*, and the great length and density of its fur, causes it to appear much stouter in its proportions ; its colouring is darker, the tips of the hairs being much more narrowly annulated with yellow than in that species. The very small size of the ears will serve to distinguish the present animal from its congeners—*Mus longipilis*, M. Renggeri, M. arenicola, &c.

The molar teeth of the upper jaw are figured in Plate 34. fig. 8, *a* ; and fig. 8, *b*, represents the middle and last molars of the lower jaw.

" Inhabited a very small island, covered with thick forest, in the central part of the Chonos Archipelago."—D.

A mouse obtained on the islets adjoining the east coast of Chiloe (where Mr. Darwin says it was common) differs from the above in being a little smaller, the tail is rather longer, and the ears are a trifle larger. In the feet, claws, colouring and character of the fur it agrees, and likewise in the pale colour and slenderness of the incisors. Its dimensions are as follows :—

	In.	Lines.		In.	Lines.
Length from nose to root of tail . .	4	0	Length from nose to ear . . .	0	10½
of tail	3	0	of ear	0	4
of tarsus (claws included) . .	0	10			

I have not the means of satisfying myself whether this be a distinct species or not ; but I think it is not.

1. Mus Renageri
2. ——— obscurus

" The nature of the country where this specimen was procured is nearly the same as in that part of the Chonos Archipelago, 150 miles to the south, where the first was obtained." D.

14. Mus Renggeri.

PLATE XV.—Fig. 1.

Mus olivaceus, *Waterh.*, Proceedings of the Zoological Society of London, for February 1838, p. 16.

M. corpore suprà subolivaceo, subtùs cinerescente ; auribus mediocribus, rotundatis, pilis parvulis fuscescentibus obsitis ; caudâ corpore breviore, pilosâ, suprà fuscâ subtùs albescente ; pedibus pilis fuscescentibus tectis.

DESCRIPTION.—Fur moderate ; ears moderate ; tail shorter than the body ; general colour gray washed with yellow ; under parts grayish white. On the upper parts and sides of the head and body the hairs are gray, broadly annulated with yellow near the apex, and dusky at the apex ; the mixture producing a yellowish gray tint, approaching somewhat towards olive :—the hairs on the under parts of the body and throat are deep gray at the base, and white at the apex ; the hairs of the feet are brownish white. The tail is tolerably well clothed with hairs ; those on the upper surface are brown, and those on the under are dirty white. The ears are well clothed, both externally and internally, with hairs of the same colour as those on the upper parts of the body. The hairs of the moustaches are for the most part whitish, and black at the base. The upper incisors are pale yellow, and the lower incisors are yellowish white.

	In.	Lines.			In.	Lines.
Length from nose to the root of tail	5	1	Length of tarsus (claws included)		0	11
of tail	2	8	of ear		0	5
from nose to base of ears	1	2				

Habitat, Valparaiso (*August and September,*) Coquimbo (*May.*)

Subsequent to the description of this species, under the name of *M. olivaceus* in the Zoological Society's Proceedings, I have imagined that perhaps that name might mislead as regards the colouring of the animal ;—it certainly has a slight olive hue, but it is not very evident. I have therefore changed the name, and substituted that of the author of the " Naturgeschichte der Säugethiere von Paraguay," &c.

In the collection there are three specimens of the present species; in one of these the hairs of the upper part and sides of the body are annulated with yellowish white, instead of yellow; hence the general hue of these parts is nearly gray.

Mus Renggeri is larger than *Mus Musculus,* and much stouter in its proportions; the fur is shorter, much less dense, and less soft than in *Mus brachiotis.*

" It inhabits dry stony places, where only a few thickets grow."—D.

15. Mus obscurus.

Plate XV.—Fig. 2.

Mus obscurus, *Waterh.,* Proceedings of the Zoological Society of London for February 1837, p. 16.

*M. suprà fusco-nigrescens, subtùs flavescens ; pedibus obscurè fuscis ; unguibus longius-
culis ; auribus mediocribus ; caudà corpore breviore, suprà nigrescente, subtùs sordidè
albâ ; vellere mediocri, molli.*

Description.—Head large; ears moderate; tail shorter than the body; fur rather
    long and glossy; the general hue of that of the upper parts and sides of the
    head and body is blackish brown, and that of the under parts is dirty yel-
    lowish white. The hairs on the upper parts are of a deep lead colour at the
    base, black at the apex, and narrowly annulated with dark yellow near the
    apex ; those of the throat and belly are lead colour at the base and yellowish
    at the tip ; the chin is white : around the eye, and on the lower part of the
    cheeks a deep yellow tint prevails. The ears are well clothed with hairs both
    externally and internally, and these are for the most part of a deep brown
    colour, as are also the hairs which cover the feet. The tail is well clothed
    with hairs, those on the upper surface are black, and those on the under are
    dirty white. Both upper and lower incisors are yellow, but the lower are
    paler than the upper.

	In.	Lines.		In.	Lines.
Length from nose to root of tail .	. 5	3	Length of tarsus (claws included) .	. 0	11½
of tail	. 2	7	of ear	. 0	4
from nose to ears . .	. 1	2½			

Habitat, Maldonado, La Plata, *(June.)*

The present species, like the foregoing, is much stouter than the common

mouse (*Mus Musculus*), its colour is much darker. In possessing a glossy fur it differs from most of its congeners ; its head is also proportionately larger, and the incisors are much stronger.

The molars of the upper jaw are figured in plate 34, fig. 9, *a*,—and fig. 9, *b*, represents those of the under jaw.

" Very abundant in gardens and hedges, far from houses ; and was easily caught in traps baited either with cheese or meat."—D.

16. Mus xanthorhinus.

Plate XVII.—Fig. 1.

Mus xanthorhinus, *Waterh.*, Proceedings of the Zoological Society of London, for January 1837, p. 17.

M. suprà fuscus flavo lavatus ; subtùs albus; rhinario flavo ; auribus parvulis, intùs pilis flavis obsitis ; mystacibus longis, canis, ad basin nigrescentibus : caudâ corpore breviore, suprà fuscâ, ad latera flavescente, subtùs sordidè albâ : pedibus anticis, tarsisque flavis, digitis albis : vellere longo, molli.

Description.—Fur moderately long and loose ; ears rather small ; tail shorter than the body ; general colour gray washed with yellow, the yellow colour prevailing, especially on the sides of the body ; muzzle, inner side of ears, and tarsus, of a rich yellow colour ; toes, chin, throat, under parts of body, and rump, white ; all the fur deep gray at the base ; the hairs on the upper parts and sides of the body broadly annulated near the apex with rich yellow, and at the apex dusky ; on the under parts of the body the hairs are broadly tipped with white. Tail rather sparingly furnished with hair, that on the upper surface brown, on the sides yellow, and on the under surface whitish. The hairs of the moustaches are white—some of them dusky at the base. The incisor teeth are rather slender, and of a pale yellow colour.

	In.	Lines.			In.	Lines.
Length from nose to root of tail .	. 3	6	Length of tarsus (claws included) .	. 0	9
of tail	. 1	7½	of ear	. 0	3¾
from nose to ear .	. 0	10			

Habitat, Hardy Peninsula, Tierra del Fuego, (*February.*)

The white, which is usually confined to the under parts of the body, in this species extends slightly on the sides of the body, and the lower portion of the cheeks.

"This species was caught on the mountains, thickly covered with peat, of Hardy Peninsula, which forms the extreme southern point of Tierra del Fuego."—D.

17. Mus canescens.

Mus canescens, *Waterh.*, Proceedings of the Zoological Society of London for February, 1837, p. 17.

M. suprà canescens, subtùs albus ; oculis flavido cinctis ; auribus parvulis, pilis pallidè flavis et plumbeis obsitis ; mystacibus mediocribus, canis, ad basin nigricantibus ; caudâ vix corpore breviore, suprà fusco-nigrâ, subtùs sordidè albâ ; pedibus anticis tarsisque flavescentibus.

DESCRIPTION.—Fur moderately long and loose ; ears small ; tail nearly equal to the body in length : general colour gray, with a wash of very pale yellow; chin, throat, and under parts of the body, white. Tail tolerably well clothed with hairs, those on the upper surface brown, and those on the under, whitish ; on the sides are some yellowish hairs. Ears with yellow hairs on the inner side; tarsi pale yellow, toes white ; muzzle and around the eye yellowish.

	In.	Lines.		In.	Lines.
Length from nose to root of tail	3	6*	Length of tarsus (claws included)	0	9½
of tail	2	1	of ear	0	4
from nose to ear	1	1			

Habitat, Santa Cruz and Port Desire, (*December.*)

" Very common in long dry grass in the valleys of Port Desire."—D.

The skull is figured in Plate 33, fig. 5, *c.* Fig. 5, *a.* represents the molars of the upper jaw ; fig. 5, *b.* those of the under jaw, and fig. 5, *d.* represents the posterior molar of the under jaw when more worn.

It was with some hesitation that I described this as a distinct species in the Society's Proceedings. I have now re-examined the specimens, and still am

---

* The dimensions given in the Proceedings of the Zoological Society were taken from a younger specimen than those here described, and there is an error in the length of the tail there given, which should be 1—10 instead of 2—10.

Fig.1 Mus xantherhinus 2 Mus nasutus.

unable to satisfy myself whether they are varieties of *Mus xanthorhinus* or not. Both of *Mus canescens* and of *Mus xanthorhinus*, I have before me what I imagine to be an adult and a young specimen. The adult and the young of *M. xanthorhinus* agree in being of a *yellowish-brown* colour, and in having the muzzle and tarsi deep yellow; both specimens of *Mus canescens* are of a *gray* colour, with an indistinct yellow wash, the muzzle and tarsi being tinted with yellow, as in *M. xanthorhinus*. Besides this difference in tint, which, perhaps, is unimportant, *M. canescens* differs from *M. xanthorhinus* in having the head larger, the tail rather longer, and the fur less soft. The specimens of this animal are both from Patagonia ; one of the specimens of *Mus xanthorhinus* was brought by Mr. Darwin from Terra del Fuego ; and as the other formed part of Captain King's collection, it in all probability came from the same locality. As I only possess one skull, I cannot speak with certainty as regards the size of the head ; the difference, however, in the stuffed specimens is considerable, and it is strange that each of the pairs should agree so perfectly, supposing the difference to be the work of the stuffer's hands.

18. Mus longipilis.

Plate XVI.

Mus longipilis, *Waterh.*, Proceedings of the Zoological Society of London for February 1837, p. 16.

M. suprà obscurè griseus, flavo lavatus ; subtùs griseus ; pedibus fuscis, unguibus longiusculis ; auribus mediocribus ; caudà corpore breviore, suprà nigrescente, subtùs fuscescente ; rhinario sub-producto : vellere longissimo, molli.

Description.—Fur very soft and silky, and extremely long—the ordinary fur of the back measuring nearly three quarters of an inch, and the longer hairs one inch in length ; ears moderate ; tail nearly as long as the body ; muzzle much pointed ; general colour gray, washed with yellow, the under parts pale gray, or grayish white ; feet brown ; ears and tail well clothed ; the hairs on the inner side of the ears are chiefly of a yellow colour, those on the upper surface of the tail are brown black, those on the under part are dirty white ; the hairs of the back are deep gray at the base, broadly annulated with yellow near the apex, and dusky at the apex ; the longer hairs are grayish black ; the hairs of the moustaches are dusky at the base, and whitish beyond that part ; the claws are long, and but slightly curved ; the incisors are slender; those on the upper jaw are yellow, and those of the under yellow-white.

	In.	Lines.				In.	Lines.
Length from nose to root of tail	. . 5	4		Length of tarsus (claws included) .	. . 1	0½
of tail	. 3	4		of ear	. . 0	6¼
from nose to ear	. . 1	2				

Habitat, Coquimbo, Chile, (*May.*)

This mouse is remarkable for the great length and softness of its fur, even among the species here described, most of which have very loose, long and soft fur.

The molars of the upper jaw are figured in Plate 33, fig. 6, *b.*—molars of the lower jaw, fig. 6, *a.*

"Inhabits dry stony places, which character of country is general in this part of Chile."—D.

19. Mus nasutus.

Plate XVII.—Fig. 2.

Mus nasutus, *Waterh.*, Proceedings of the Zoological Society of London for February 1837, p. 16.

M. suprà obscurè flavescenti-fuscus, ad latera fulvescens; subtùs obscurè fulvo tinctus: pedibus pilis obscurè fuscis tectis; unguibus longis; auribus mediocribus; caudâ corpore breviore, suprà fuscâ, subtùs sordidè albâ: rhinario producto.

DESCRIPTION.—Muzzle very long and pointed, ears small, tail shorter than the body, claws long and but slightly arched; inner, rudimentary toe of the fore foot furnished with a pointed claw; fur moderate, and slightly glossy: general colour yellowish brown, of the sides of the body yellow, of the under parts pale yellow; the chin, throat and chest whitish: feet brown; ears well clothed with hairs, those on the inner side are most of them yellow, but some are black. All the fur is of a deep lead colour at the base; the hairs on the upper parts and sides of the head and body are broadly annulated with deep golden yellow near the apex, and blackish at the apex; on the upper parts long brownish black hairs are thickly interspersed with the ordinary fur, but on the side of the body they are less numerous, hence on this part

Phascogale.

the yellow tint prevails; on the under parts of the body the hairs are broadly tipped with pale yellow, and in parts with white: the tail is but sparingly clothed with hairs, those on the upper surface are of a dark brown colour, and those on the under are pale brown. The incisors are very slender and of a very pale yellow colour.

	In.	Lines.		In.	Lines.
Length from nose to root of tail .	. 5	2	Length of tarsus (claws included) .	. 1	0½
of tail	. 2	8	of ear	. 0	5
from nose to ear .	. 1	3			

Habitat, Maldonado, La Plata, (*June*.)

The specific name *nasutus* has been applied to this mouse on account of its elongated and slender muzzle*, the tip of which extends nearly 4 lines beyond the upper pair of incisors: the rudimentary toe of the fore foot, instead of having the usual rounded nail, has a short pointed claw. Its fur is not so soft, nor yet so long as in many of the preceding species, and there is a greater admixture of yellow in its colouring. The claws appear to be adapted to burrowing.

The skull (which is not quite perfect) is figured in Plate 33, fig. 7, *a*, its length is 1 in. 3 lines. Fig. 7, *b*, represents the molars of the upper jaw, and fig. 7, *c*, those of the under jaw. The lower jaw, which is of a very slender and elongated form, is figured in Plate 34, fig. 10, *a*.

"Was caught in a small thicket on an open grassy plain, by a trap baited with a piece of bird. This mouse when alive possesses a marked character in the extreme acumination of its nose."—D.

20. MUS TUMIDUS.

PLATE XVIII.

Mus tumidus, *Waterh.*, Proceedings of the Zoological Society of London for February 1837, p. 15.

M. brunneus, nigro lavatus; rostro ad apicem, labiis, mento, gulâ, pectore, abdomineque albis; naso suprà nigrescente; auribus mediocribus rotundatis; corpore crasso; caudâ capite corporeque breviore, pilis nigricantibus, subtùs albescentibus prope basin, vestitâ; artubus pedibusque grisescentibus; vellere longo, molli; unguibus longis.

DESCRIPTION.—Body stout; head large; tail nearly as long as the head and body;

* In *Mus longipilis* and *M. brachiotis* may be perceived an approach to this elongated form of the muzzle.

I

inner toe of the fore foot with a distinct, pointed claw; claws rather large,
those of the fore feet but slightly arched. Fur rather long, and moderately
soft; general tint of the upper parts of the body, brown, of the sides of the
head and body, grayish, but with a yellow wash; the lower part of the sides
of the body and of the cheeks, the tip of the muzzle, and the whole of the
under parts, white; feet dirty white; ears densely clothed with short hairs,
those on the inner side chiefly of an ashy-brown colour, and those on the
outer side dusky; the hairs of the back are of a deep lead colour at the base,
black at the tip, and annulated with yellow near the tip; the longer hairs,
which are thickly interspersed, are totally black; on the under parts of the
body the hairs are gray at the base, and broadly tipped with white; the upper
surface of the muzzle is blackish; the moustaches are black; the incisors
are yellow.

	In.	Lines.		In.	Lines.
Length from nose to the root of tail	6	9	Length of tarsus (claws included) .	1	6
of tail	5	4	of ear	0	7
from nose to ears . .	1	8			

Habitat, Maldonado, La Plata (*June*.)

This species is about the size of *Mus Rattus*, but is stouter in its proportions;
as in *Mus nasutus*, the thumb is furnished with a pointed claw. The molars of the
lower jaw are figured in Plate 34, fig. 11, *a*.

"This rat was caught in so wet a place amongst the flags bordering a lake,
that it must certainly be partly aquatic in its habits."—D.

21. Mus Braziliensis.

Plate XIX.

Rat du Brézil, *Geoff.*

*M. suprà fuscus fulvo lavatus ; lateribus capitis corporisque æquè ac abdomine auratis ;
gulâ pectoreque albis ; pedibus pilis sordidè flavis tectis ; auribus parvulis ; caudâ
caput corpusque ferè æquante ; vellere longo, molli.*

DESCRIPTION.—Head somewhat arched, and rather short; ears small; tail about
equal in length to the head and body, measured in a straight line; tarsi
large. Fur long, and rather soft; general colour deep golden yellow: on
the upper surface of the head and the back, long glossy black hairs are
thickly interspersed, and produce, with the admixture of the deep golden

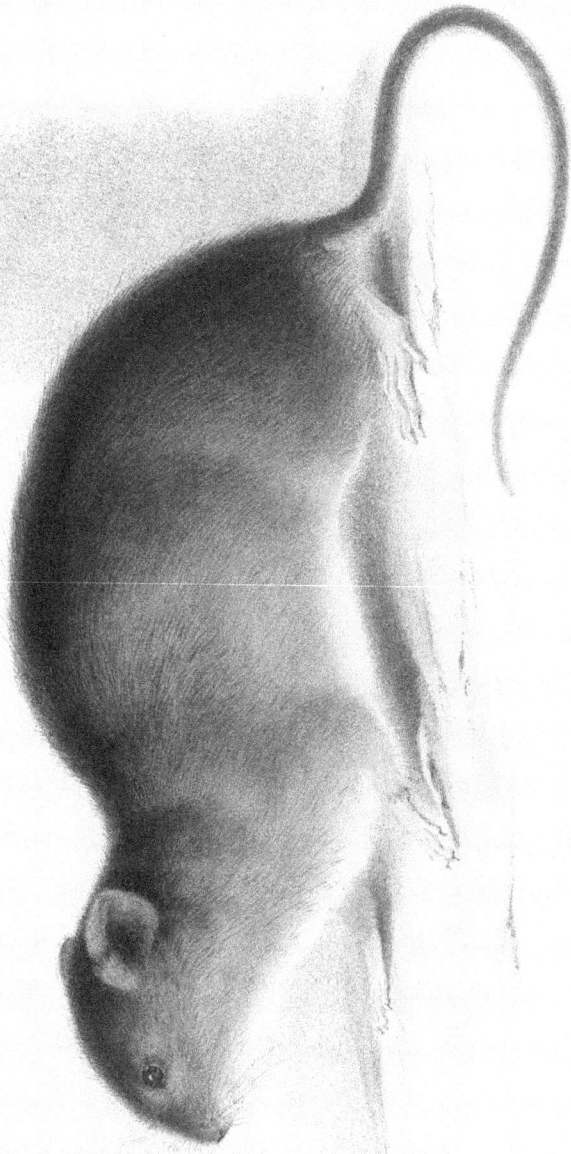

colour of the ordinary fur, a dark brown tint; chin, throat, chest, and rump, white; the hairs covering the upper surface of the feet are of a dirty yellowish-white colour, and on the toes nearly white : ears densely clothed with longish hairs, those on the inner side chiefly of a deep golden colour, and those on the outer side brownish; the ears are partially hidden by the long fur of the head; tail sparingly clothed with hairs, above brown, and beneath brownish-white : the fur of the back is of a deep gray colour at the base, annulated with deep golden yellow near the apex, and blackish at the apex; the longer hairs are black; the hairs of the belly are pale gray at the base, and broadly tipped with golden yellow colour; the white hairs on the throat, chest, and rump are of an uniform colour—not tinted with gray at the root;—the hairs of the moustaches are black : the incisors of the upper jaw are of a deep orange colour, and those of the lower jaw are yellow: the thumb nail is truncated.

	In.	Lines.							In.	Lines.
Length from nose to root of tail	. . 8	6	Length of tarsus	.	.	.	.	.	2	0
of tail	. 7	9	of ear	.	.	.	.	.	0	6½
from nose to ear	. . 1	8								

Habitat, Bahia Blanca, (*September.*)

This species is nearly equal in size to the common rat (*Mus decumanus*). Of its skull * I possess but the anterior portion (see Pl. 33. fig. 3, *a*. and 3, *b*.) : it appears to have been about the same size as that of *M. decumanus*, its proportions, however, are different: the nasal portion is broader and shorter, the ant-orbital outlet is rather smaller; the plate, forming the anterior root of the zygomatic arch, and which protects this outlet, has its anterior edge distinctly emarginated, and not nearly straight as in *M. decumanus*,—the zygomatic arch is stouter, the space between the orbits is narrower, the palate is more contracted, the incisors are much broader, less deep from front to back, and have the anterior surface more convex ; the molar teeth are larger ; the lower jaw (see Plate 34. fig. 12, *a*.) when compared with that of *Mus decumanus* also offers many points of dissimilarity ; the principal differences consist in its greater strength, the comparatively large size and breadth of the articular surface of the condyles, the upright position of the coronoid process—a perpendicular line dropt from the apex of which would touch the posterior part of the last molar—and the great

---

* I am sorry to say the artist has not drawn this skull with his usual fidelity, a circumstance which I did not perceive until it was too late to make any alteration: it is too large, and the incisors are represented as projecting forwards too much ; they are in the original so nearly at right angles with the upper surface of the skull that but a very small portion of them is seen, when it is viewed, as represented at fig. 3, *a*.

extent of the *symphysis menti*. In the form of the incisors, the more contracted palate, the great extent of the *symphysis menti*, and in fact in most of the points of dissimilarity, between the skull of the present animal and that of *Mus decumanus*, here pointed out, it will be perceived, there is an approach made to the *Arvicolidæ*.

The dimensions of the skull (so far as an imperfect specimen will allow of their being taken) are as follows :—

	In.	Lines.
Distance between front of incisors, (upper jaw) and the first molar tooth	0	8
Longitudinal extent of the three molars on either side, taken together .	0	4¼
Length of nasal bones	0	7¼
——— of incisive *foramina*	0	4¼
Width between orbits	0	2½
Length of *ramus* of lower jaw	1	1¼

Fig. 3, *c*, Plate 33, represents the molar teeth of the upper jaw. Fig. 3, *d*, those of the upper jaw.

" This rat was caught at Bahia Blanca where the plains of Patagonia begin to blend into the more fertile region of the Pampas. It lived in holes amongst the tussocks of rushes, on the borders of a small, still, brook ; in its manner of diving and aquatic habits it closely resembled the English water-rat, (*Arvicola amphibia*.)"—D.

When at Paris I examined what I believe to be the original *Mus Braziliensis*, since the specimen was labelled " *Rat de Brazil St. Hilaire*, 1818." It agrees perfectly with the present animal excepting in being rather smaller, the length from the nose to the tail being 7 inches and 4 lines—the length of the tail is 7 inches 9 lines, and that of the tarsus is 1 inch 11 lines ; this difference in the length of the body may arise from difference of age, or even of sex. In the Paris Museum I saw what appeared to me to be a variety of the same species in which the under parts of the body are white.

I have been minute in my description of the *Mus Braziliensis*, since it is confounded by Desmarest, Fischer and Lesson with the *Rat troisieme* or *Rat Angouya* of Azara, which I believe to be a very different animal. The description given by the authors just mentioned are taken from Azara, who gives the following characters to distinguish the Rat Angouya : " Du museau à la queue, et sur les côtés du corps tout est brun-cannelle, parceque les poils ont une petite pointe cannelle ; puis, ils sont obscurs et enfin blanc vers las peau. Toute la partie inférieure de l'animal est blanchâtre, plus claire sous la tête, et plus foncée entre les jambes de devant ; le pelage est doux, très-serré, et le poil, qui est à la racine de l'oreille, cache le conduit de celle-ci."

Mus tomaris

	In.	Lines.		In.	Lines.
Length from nose to root of tail (English			Length of ears	0	9¾
measure)	0	0	of tarsus (the claws included) .	1	3¼
of tail	6	6½			

It appears from this description that the *Mus Angouya* is a smaller animal, and differs both in colouring and proportions from the *Mus Braziliensis*. Brandt has figured and described a rat under the name of *Mus Angouya*, which in many respects agrees better with Azara's description ; there are, however, discrepancies in the dimensions.

22. Mus micropus.

PLATE XX.

Mus micropus, *Waterh.*, Proceedings of the Zoological Society of London for February 1837, p. 17.

M. suprà fuscus ; subtùs cinerescenti-albus, pallidè flavo tinctus ; pedibus pilis sordidè albis tectis, antipedibus parvulis ; auribus parvulis ; caudâ, quoad longitudinem, corpus ferè æquante, suprà fuscâ, subtùs sordidè albâ.

DESCRIPTION.—Form stout, ears rather small, tail nearly equal to the body in length, fur very long and moderately soft, general colour of the upper parts of head and body, brown ; of the sides of the body grayish, faintly washed with yellow, of the under parts grayish white, faintly tinted with yellow; hair covering the upper surface of the feet dirty white; on the tarsus there is a very slight yellow tint ; ears well clothed with hairs, those on the inner side chiefly of a yellow colour; tail above, dusky brown; beneath dirty white : hairs of moustaches black at the base and grayish at the apex ; incisors pale yellow : hairs of the back deep gray at the base, annulated with brownish yellow near the apex, and dusky at the apex ; longer hairs dusky black ; hairs of the belly deep gray at the base and broadly tipped with yellowish white.

	In.	Lines.		In.	Lines.
Length from nose to root of tail .	6	0	Length of tarsus (claws included) .	1	0¾
of tail	3	8	of ear	0	6
from nose to ear . .	1	4			

Habitat, Santa Cruz, Patagonia, (*April.*)

The molars of the upper jaw are figured in Plate 34, fig. 13, *a*, and those of the lower jaw, fig. 13, *b*.

"Caught in the interior plains of Patagonia in lat. 50°, near the banks of the Santa Cruz."—D.

23. Mus griseo-flavus.

Plate XXI.

Mus griseo-flavus, *Waterh.*, Proceedings of the Zoological Society of London for February 1837, p. 28.

M. suprà griseus flavo-lavatus, ad latera flavus, subtùs albus ; pedibus albis ; auribus magnis et ferè nudis ; caudâ caput corpusque ferè æquante, suprà fusco-nigricante, subtùs albâ ; vellere longo, molli.

Description.—Ears large ; tail rather shorter than the head and body taken together ; tarsi slender, and moderately long ; fur long and very soft ; general tint of the upper parts of head and body grayish, washed with brownish yellow ; on the sides of the body a palish yellow tint prevails ; feet, chin, throat, and under parts of body pure white ; tail rather sparingly clothed with hairs, those on the apical portion rather long, and forming a slight pencil at the tip ; on the upper side and at the tip of the tail the hairs are brown, on the under side they are dirty white ; the ears are very sparingly clothed with minute brownish yellow hairs internally ; externally, on the fore part, the hairs are rather longer and of a brown colour ; the upper incisors are orange, and the lower incisors are yellow ; the hairs of the moustaches are long, and of a black colour ; the hairs of the back are deep gray at the base, brownish at the tip, and annulated with pale brownish yellow near the tip ; the longer hairs are brown ; the hairs of the belly are white externally, and gray at the base ; on the throat the hairs are white to the root.

	In.	Lines.		In.	Lines.
Length from nose to root of tail	6	8	Length of tarsus (claws included)	1	2½
of tail	5	6	of ear	0	8
from nose to ear	1	4½			

Habitat, Northern Patagonia (*August.*)

The molars of the upper jaw are figured in Plate 34, fig. 15, *a*, and those of the lower jaw, fig. 15, *b*.

"Inhabits the dry gravelly plain, bordering the Rio Negro."—D.

Mus arboreus flavus

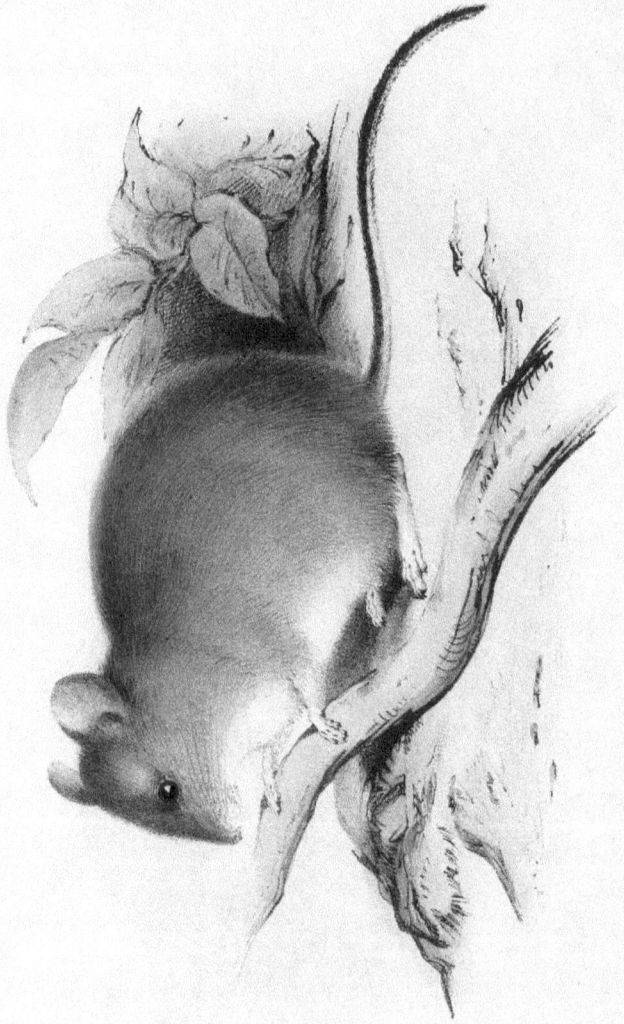

Mus xanthopygus

24. Mus xanthopygus.

Plate XXII.

Mus xanthopygus, *Waterh.*, Proceedings of the Zoological Society of London for February 1887, p. 28.

M. suprà pallidè brunneus flavo-lavatus, ad latera flavescens, subtùs albus ; capite griscescente; natibus flavis ; pedibus albis ; auribus majusculis pilis, albis et flavis intermixtis obsitis ; caudâ quoad longitudinem, corpus ferè æquante, suprà nigricante, subtùs albâ ; vellere longo et molli ; mystacibus perlongis albescentibus, ad basin nigris.

Description.—Ears rather large, tail rather longer than the body, tarsi moderately long and somewhat slender: fur long and very soft : prevailing tint pale yellow; on the back there is a brownish hue owing to the long hairs, which are thickly interspersed with ordinary fur, being of that colour: in the region of the tail the hairs are of a rich yellow colour ; the tip of the muzzle is white, the feet, chin, throat and the whole under parts of the body are white ; on the chest and belly a faint yellowish hue is observable : the tail is well clothed with tolerably long hairs, those on the apical portion are the longer, on the upper side of the tail they are of a brown colour, and on the under side they are pure white : the ears are well clothed with tolerably long hairs, those on the inner side are of a pale yellowish colour, externally on the fore part they are brown, and on the hinder part they are yellowish white: the hairs of the moustaches are numerous and very long; some of them are white, but the greater portion are brownish black at the base and whitish at the apex: the upper incisors are yellow, and the lower are yellow-white: the hairs of the ordinary fur on the back are gray at the base, brownish at the tip, and very pale yellow near the tip : the hairs on the belly are gray at the base and white externally.

	In.	Lines.		In.	Lines.
Length from nose to root of tail	5	3	Length of tarsus (claws included)	1	1
of tail	3	10	of ear	0	7
from nose to ear	1	3			

There are three specimens of the present species in Mr. Darwin's collection ; two of them were caught when shedding their fur, and having lost the longer black hairs, have the upper parts of the body of a paler colour; their general tint is very pale, and may be described as gray, with a wash of pale yellow.

This species is closely allied to the last, but differs in being rather smaller, in having smaller ears which are well clothed with hair, and not sparingly furnished as in *Mus griseo-flavus*, and in having a shorter tail which, like the ears, is more densely clothed with hairs ; in the structure of the molar teeth there also differences which will be better understood by comparing the drawings. Fig. 16, *a*, Plate 34, represents the molars of the upper jaw, and 16, *b*, those of the lower jaw.

"Extremely abundant in the coarse grass and thickets in the ravines at Port Desire and Santa Cruz : was caught in a trap baited with cheese."—D.

25. Mus Darwinii.

PLATE XXIII.

Mus Darwinii, *Waterh.*, Proceedings of the Zoological Society of London for February 1837, p. 28.

M. suprà pilis pallidè cinnamomeis et nigrescentibus intermixtis ; ante oculos cinerascentibus ; genis, lateribus corporis, et caudâ prope basin, pallidè cinnamomeis ; partibus inferioribus pedibusque albis; auribus permagnis ; caudâ caput corpusque ferè æquante, suprà fusco-nigricante, subtùs albâ.

DESCRIPTION.—Form robust ; ears immensely large ; tail nearly equal in length to the head and body taken together ; fore feet very small ; tarsi moderate ; fur very long and soft ; general tint of the upper parts pale cinnamon yellow ; on the rump a richer yellow hue prevails, and on the back there is a brownish tint, owing to the interspersed long hairs being of that colour ; the upper surface of the head is grayish ; the cheeks, like the sides of the body, are of a delicate yellow colour, faintly clouded with brown ; the sides of the muzzle, lower part of the cheeks and sides of the body, and the whole under parts, are pure white ; the feet and tail are also white, if we except the upper surface of the latter, which is dark brown ; the yellow tint of the sides of the body is extended downwards on the outer side of the fore legs and on the back of the hinder legs ; the ears are but sparingly furnished with hair, excepting on the fore part, externally, where they are of a brownish colour ; the minute hairs which cover the remaining parts of the ear are very pale ; the tail is well clothed with hairs ; the hairs of the moustaches are numerous and very long ; they are for the most part blackish at the base, and gray at the apex ; the incisors are rather slender, the upper pair are an orange colour, and the lower, yellow ; the hairs of the ordinary fur of the back are gray at

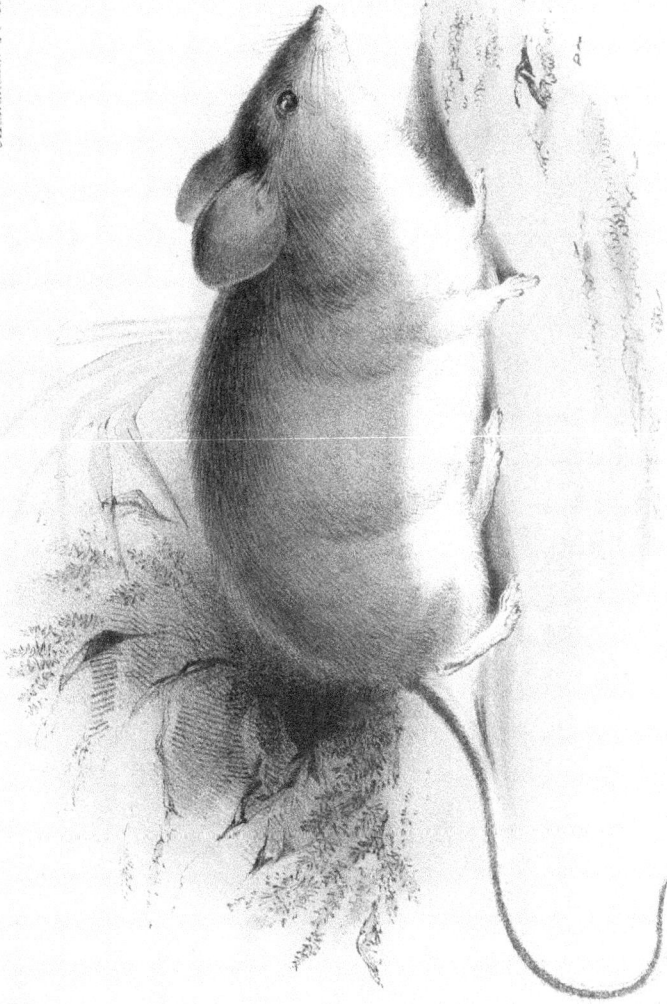

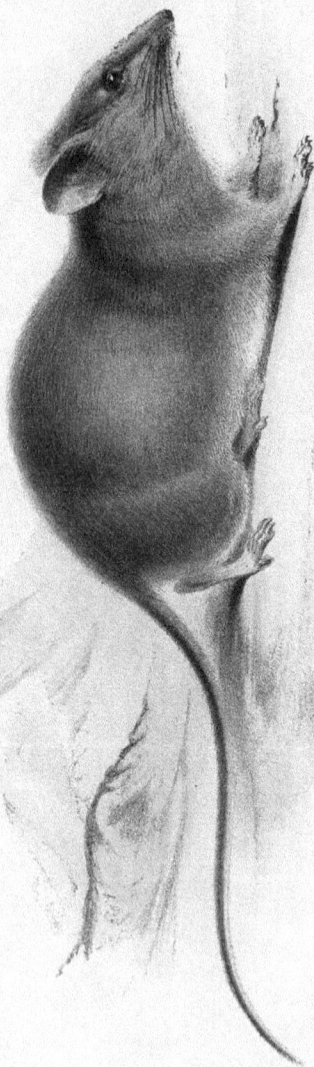

Mus Galapagoensis

the base, broadly annulated with pale cinnamon yellow near the apex, and brownish at the apex ; the hairs of the belly are deep-gray at the base, and white externally, those on the throat are pale gray at the base.

	In.	Lines.		In.	Lines.
Length from nose to root of tail .	. 6	0	Length of tarsus (claws included) .	. 1	1½
of tail	. 4	9	of ear	. 0	11¾
from nose to ear .	. 1	4½	Width of ear	. 1	0½*

Habitat, Coquimbo, Chile, (*May*.)

This species is evidently allied to the two preceding ; and perhaps the " Rat quatrieme, ou Rat oreillard" of Azara, (*Mus auritus*, Desm.) will form one of this little group. The molar teeth of the upper jaw are figured in Plate 34, fig. 17, *a*— those of the lower jaw, fig. 17, *b*.

" Inhabits dry stony places."—D.

26. Mus Galapagoensis.

Plate XXIV.

M. suprà fuscus, flavo-lavatus, ad latera flavescens, subtùs albus : pedibus pilis sordidè albis tectis : auribus mediocribus ; caudá, quoad longitudinem, caput corpusque ferè æquante : vellere longo.

Description.—Ears moderate, slightly pointed ; tarsi moderate ; tail slender, nearly as long as the head and body ; fur long, and not very soft ; upper parts of the body of a brownish hue, a tint produced by the admixture of black and palish yellow hairs ; on the sides of the body the longer black hairs are less abundant, and the prevailing colour is yellow ; under parts of the body white, with a very faint yellow tint ; feet furnished above with dirty white hairs ; ears rather sparingly clothed with hairs, those on the inner side of a yellow colour, and those on the outer side dusky ; tail above brown, and beneath whitish ; the hairs of the moustaches black ; the incisors deep yellow ; the hairs on the back are deep gray at the base, broadly annulated with palish yellow near the apex, and blackish at the apex ; the longer hairs black ; on the belly the hairs are gray at the base, and broadly tipped with yellowish white.

<hr>

* It is not easy to measure the *width* of the ears in these animals : upon measuring with a thread over the curve of the outer side I have found the width of the ears of the present animal to be as above given,—the dimension slightly exceeding that stated in the Proceedings of the Zool. Soc.

	In.	Lines.			In.	Lines.
Length from nose to root of tail	6	0	Length of tarsus (claws included)		1	2
of tail	4	9	of ear		0	7
from nose to ear	1	3¾				

Habitat, Chatham Island, Galapagos Archipelago, Pacific Ocean, (*October.*)

This species is less than *Mus Rattus.* The upper parts of the body have a slightly variegated appearance.

The skull of *Mus Galapagoensis* (Plate 33, fig. 8, *a,*) is rather smaller than that of *M. Rattus*, the nasal portion is proportionately longer, the cranial shorter, and the interparietal bone is smaller, especially in antero-posterior extent; its length is 15 lines, and its breadth is 8¼ lines. The lower jaw is figured in Plate 34, fig. 14, *a.* Fig. 8, *b,* of Plate 33, represents the molars of the upper jaw, and fig. 8, *c,* those of the lower jaw.

" This mouse or rat is abundant in Chatham Island, one of the Galapagos Archipelago. I could not find it on any other island of the group. It frequents the bushes, which sparingly cover the rugged streams of basaltic lava, near the coast, where there is no fresh water, and where the land is extremely sterile."—D.

27. Mus Fuscipes.

Plate XXV.

M. suprà fusco-nigrescens, subtùs griseus; pedibus fuscis; auribus mediocribus, caudá, quoad longitudinem, caput corpusque ferè æquante: vellere longissimo, molli.

Description.—Form stout; ears moderate; tail equal to the body in length; tarsi moderate; fur very long. General tint of the upper part and sides of the head and body blackish brown with an admixture of gray; of the under parts grayish white; feet brown, the hairs grayish at the tip: tail black and but sparingly clothed with short bristly hairs: ears rather sparingly clothed with hairs, which are for the most part of a brownish gray colour. The ordinary fur of the back is about ¾ of an inch in length and very soft—of a deep gray colour, broadly annulated with brownish yellow near the tip and blackish at the tip: the longer hairs which are black, measure upwards of 1¼ inches in length. The upper incisors are of an orange colour and the lower are black.

	In.	Lines.			In.	Lines.
Length from nose to root of tail	6	6	Length of ear		0	6½
of tail	4	3	of tarsus (claws included)		1	1
from nose to ear	1	6				

Habitat, Australia, King George's Sound, (*March.*)

Mammalia not belonging to the order *Marsupiata* are rare in the Continent of Australia. Besides the Dog, we are acquainted with none excepting a few species of Rodents, and these all belong to the family *Muridæ*.

The present animal adds one to the limited number already known: in the Museum of the Zoological Society there is another species, the characters of which I will point out in the next description.

Mus fuscipes is remarkable for the great length and softness of its fur, and the brown colour of its feet: it is rather less than *Mus Rattus*, and of a stouter form. Not having had an opportunity of examining the molar teeth and the cranium of this animal, I cannot be positive that it is a species of the genus *Mus;* in external characters and the form of the incisor teeth, however, it agrees perfectly with the animals of that genus.

"This animal was caught in a trap baited with cheese, amongst the bushes at King George's Sound."—D.

28. Mus Gouldii.

M. vellere longo, molli, ochraceo, pilis nigricantibus adsperso, his ad latera rariori-bus: corpore subtùs, pedibusque albis: auribus majusculis: caudâ, capite corporeque paulo breviore.

DESCRIPTION.—Ears rather large and slightly pointed, tarsi slender and tolerably long; tail about equal in length to the body and half the head; fur long and soft; general colour pale ochreous yellow; on the back there are numerous long black hairs interspersed with the ordinary fur, which gives a darker hue and somewhat variegated appearance to that part; feet, chin, throat, and the whole under-parts of the body white; ears brown, sparingly clothed with minute yellow hairs, both externally (excepting on the forepart, where they are brownish) and internally; tail brownish above, and yellowish white beneath; the hairs of the moustaches long, and of a brown colour; upper incisors deep orange, lower incisors yellow; claws white. The hair of the back is of a deep lead colour at the base, pale ochre near the apex, and dusky at the apex; the longer hairs are black; the hairs of the belly are deep gray at the base and broadly tipped with white.

	In.	Lines.		In.	Lines.
Length from nose to root of tail .	. 4	8 .	Length of tarsus (claws included)	, 1	0½
of tail . . .	. 3	6	of ear	. 0	7
from nose to ear .	. 1	0½			

·VAR. β.—General colour of the fur pale ochreous yellow, the feet, under side of the tail and the whole of the under parts, as well as the lower portion of the sides of the body, white; hairs of the back palish gray at the base, those of the belly indistinctly tinted with very pale gray at the roots; ears and moustaches pale brown.

Habitat, New South Wales.

This species is about half-way between *Mus Rattus* and *Mus musculus* in size, and is remarkable for its delicate colouring. The molar teeth are figured in Plate 34; fig. 18. *a*, represents the molars of the upper jaw, and fig. 18. *b*, those of the lower.

Genus—REITHRODON.*

Dentes primores $\frac{2}{2}$; *inferioribus acutis, gracilibus, et anticè lævibus; superioribus gracilibus, anticè longitudinalitèr sulcatis.*

Molares utrinque $\frac{4}{4}$ *radicati; primo maximo, ultimo minimo: primo superiore plicas vitreas duas externè et internè alternatim exhibente; secundo, et tertio, plicas duas externè, internè unam: primo inferiore plicas vitreas tres externè, duas internè; secundo, plicas duas externè, unam internè; tertio unam externè et internè, exhibentibus.*

Artus inæquales: antipedes 4-dactyli, cum pollice exiguo: pedes postici 5-dactyli, digitis externis et internis brevissimis.

Ungues parvuli et debiles. Tarsi subtùs pilosi.

Cauda mediocris, pilis brevibus adpressis instructa.

Caput magnum, fronte convexo: oculis magnis: auribus mediocribus.

The present genus according to my views belongs to the family *Muridæ*. The modifications of structure which have led me to separate it from the genus *Mus* are as follows:

External characters. — The most conspicuous points of distinction between the external characters of *Reithrodon* and *Mus* (if we regard *M. rattus*, *M. decumanus* or *M. musculus* as typical examples of that genus,) consist in the arched form of the head, the large size of the eyes, the stout form of the body, and the upper incisors being grooved. The ears, tail and feet are more densely

---

* Ρειθρος, a channel; Οδον, a tooth.

Reithrodon cuniculoides

clothed with hairs, and the tarsus is covered with hair beneath,—at least the hinder portion.

Cranium.—The skulls of the species of the present genus differ from those of the species of *Mus* in being proportionately shorter and broader, and more arched; the facial portion of the skull is larger, compared with the cranial, the space between the orbits is narrower, and the orbits are larger; the palate is narrower and the incisive foramina are more elongated and larger. The pterygoids approximate anteriorly, so that the posterior *nares* are greatly contracted. As in the genus *Mus* the anterior root of the zygomatic arch is directed upwards from the plane of the palate, and forwards in the form of a thin plate, protecting an opening behind, which leads into the nasal cavity, and also forming the outer boundary both of the ant-orbital foramen, and a second opening whose outlet is directed upwards. This thin plate, however, is narrower than is usually found in the genus *Mus*. The most striking differences observable in the lower jaw consist in the smaller size of the coronoid process, and its being curved outwards; the condyloid process is narrower, and the angle of the jaw, or descending ramus, approaches more nearly to a quadrate form—the posterior edge of the jaw is more deeply emarginated.

Dentition.—The incisors are narrow and compressed as in the genus *Mus*, but they are less deep from front to back; those of the upper jaw (Plate 33. fig 2. *b.*) have each a distinct longitudinal groove, which is situated nearer to the outer than to the inner edge of the tooth. Close to the inner edge of each of these teeth an indistinct second longitudinal groove may be seen by means of a lens. The lower incisors are nearly equal in width to the upper.

The crowns of the molar teeth in the young *Reithrodon* are higher than in *Mus*, and they are rootless; in the adult animal, however, they possess distinct roots. The folds of enamel form sigmoid flexures, are closely approximated to each other, and those of the opposite sides of the tooth meet.

1. Reithrodon cuniculoïdes.

Plate XXVI

Reithrodon cuniculoïdes, *Waterh.*, Proceedings of the Zoological Society of London for February 1837, p. 30.

R. suprà griseus, flavo-lavatus, pilis nigris intermixtis; abdomine gulâque pallidè flavis; natibus albis; pedibus albis; auribus mediocribus, intùs pilis flavis, extùs pilis pallidè flavis, obsitis; pone aures, notâ magnâ albescenti-flavâ; caudâ corpore breviore, suprà pallidè fuscâ, subtùs albâ.

DESCRIPTION.—Head rather large and arched; ears moderate; tail nearly as long as the body; tarsi rather long; fur long and very soft. General tint of the upper parts of the body grayish brown, with a considerable admixture of yellow; of the sides of the body grayish tinted with yellow; the lower portion of the cheeks, and the lower half of the sides of the body are of a delicate yellow colour; the under parts of the head and body are yellowish white; the fore part of the thighs is whitish; the rump, feet, and tail are white, excepting the upper surface of the latter, which is brown; behind each ear there is a patch of yellowish white hairs. The ears are tolerably well-clothed with hairs; those on the inner side are for the most part of a yellow colour, but towards the posterior margin they are brown; externally, the hairs are also yellow, excepting on the fore part, where they are dusky brown. The hairs of the moustaches are very long and numerous; black at the base, and grayish at the apex. The feet are well clothed with hairs which cover and nearly hide the claws; the under side of the tarsus is clothed with grayish brown hairs. The tail is well clothed with tolerably long hairs which completely hide the scales. The hairs on the back are of a deep gray colour at the base, broadly annulated with yellow near the apex, and dusky at the apex: the longer hairs are black: on the throat and belly the hairs are deep gray at the base, and broadly tipped with pale yellow—towards the cheeks and sides of the body with a deeper yellow. The incisors are yellow.

	In.	Lines.		In.	Lines
Length from nose to root of tail .	6	5	Length of tarsus (claws included) .	1	4¼
of tail	3	3½	of ear	0	7
from nose to ear . . .	1	4			

Habitat, Patagonia, (*April and January*).

In the arched form of the head this little animal bears considerable resemblance to a young rabbit, a resemblance which has struck almost all who have seen it, I have therefore applied to it the specific name *Cuniculoïdes*. The skull is figured in Plate 33, fig. 2. *a.*, its dimensions are as follows:—

	In.	Lines.
Total length	1	4
Width	0	10
Length of nasal bones	0	7
of incisive foramina	0	4¾
Distance between the outer surface of the incisors and the front molar upper jaw .	0	5
Longitudinal extent of the three molars of the upper jaw	0	3¾
Length of a ramus of the lower jaw, without the incisor	0	9¾

The molar teeth of the upper jaw are figured in Plate 33, fig. 2, *c.* and

2, *e*; of the lower jaw, fig. 2, *d*. Fig. 2, *b*, represents the incisors of the upper jaw magnified. Fig. 21, *a*, Plate 34, represents the skull, viewed from beneath, fig. 21, *b*, is the side view of the same, and fig. 21, *c*, is the lower jaw.

"Specimens were procured at Port Desire, St. Julian, and Santa Cruz; at this latter place they were caught in numbers, (in traps baited with cheese,) both near the coast and on the interior plains. A specimen from Santa Cruz weighed 1336 grains. In the early part of January, there were young individuals at Port St. Julian."—D.

2. REITHRODON TYPICUS.

Reithrodon typicus, *Waterh.*, Proceedings of the Zoological Society of London for February 1837, p. 30.

R. vellere suprà pilis flavescenti-fuscis et nigrescentibus intermixtis composito; regione circa oculos, genis, lateribusque corporis auratis, pilis pallidè fuscis intermixtis; partibus inferioribus auratis; rhinario ad latera flavescenti-albo; auribus magnis, intùs pilis flavis, extùs flavis et fuscis, indutis; caudâ suprà pallidè fuscâ, subtùs sordidè albâ; pedibus albis.

DESCRIPTION.—Ears large; tarsi moderate; fur moderately long; general tint of the upper parts brown—of the upper surface of the head blackish; on the cheeks and flanks a rich yellow tint prevails; the under parts of the head and body are bright yellow; the feet are white; the tail is brownish above and dirty white beneath. The ears are tolerably well clothed with hairs, and these are of a yellowish colour, excepting on the fore part, externally, where they are brown; the tarsi are covered beneath with grayish brown hairs; the hairs of the moustaches are numerous and moderately long, black at the base and grayish at the apex. The hairs of the back are deep gray at the base, broadly annulated with yellow near the apex, and black at the apex; on the upper surface of the head the hairs are very narrowly annulated with yellow, hence a blackish hue prevails. The longer hairs on the back are black; the hairs of the throat and belly are gray at the base, and broadly tipped with yellow. The incisors are yellow.

	In.	Lines.			In.	Lines.
Length from nose to root of tail	6	0	Length of tarsus (claws included)	.	1	2¼
of tail		?*	of ear	.	0	8½
from nose to ear . . .	1	4½				

Habitat, Maldonado, La Plata, (*June*).

* The tail is imperfect.

This species is of a darker colour than the last, its ears are much larger and the tarsi are shorter. It has the same rabbit-like appearance. The molar teeth of the lower jaw are figured in Plate 33, fig. 4, *a*.

" This mouse, when alive, from its very large eyes and ears, had a singular appearance, somewhat resembling that of a little rabbit. It frequents small thickets in the open grassy savannahs near Maldonado, and was caught with facility by means of traps baited with cheese. '—D.

3. REITHRODON CHINCHILLOIDES.

PLATE XXVII.

R. vellere longissimo et mollissimo; corpore suprà et ad latera cinereo, flavescenti-fusco lavato, subtùs flavescenti-albo ; caudá corpore breviore, suprà fuscá, subtùs albâ : auribus parvulis : tarsis mediocribus.

DESCRIPTION.—Ears small; tail shorter than the body; tarsus moderate; fur long and extremely soft. General hue of the upper parts of the head and body ashy-brown ; the lower part of the cheeks and sides of the body are of a delicate yellow colour; the under parts of the head and body and the rump are cream colour. The ears are blackish ;* the tail is tolerably well clothed with longish hairs, which are, however, not so thickly set as to hide the scales —on the upper side they are blackish brown ; on the sides and beneath they are white. The feet are white. All the fur is of a deep gray colour at the base; the hairs of the back are of a very pale yellow colour (almost white) near the tip, and brown at the tip; the longer hairs are black at the apex. The incisors are yellow; the hairs of the moustaches are numerous and very long—some of them are whitish, and others are black at the root, and gray at the apex.

	In.	Lines.		In.	Lines.
Length from nose to root of tail . .	5	0	Length of tarsus (claws included) .	1	0
of tail	2	4	of ear	0	5½
from nose to ear . .	1	2			

Habitat, South shore of the Strait of Magellan, near the Eastern entrance.

This little animal was preserved in spirit, and has since been mounted, it is

---

* They are naked, but I suspect the hair has been rubbed off.

probable, therefore, that the colours have been slightly changed. It is of a smaller size than either of the preceding species. Its fur is long, extremely soft, and somewhat resembles that of the Chinchilla. The ears are smaller, and the tail is shorter, and less densely clothed with hairs than in *Reithrodon cuniculoides*. The skull (see Plate 43, fig. 20, *a*, 20, *b*, and 20, *c*,) differs in many respects from that of the species last mentioned. It is of a smaller size, the nasal portion is proportionately shorter and narrower, the incisive foramina are shorter; the pterygoid processes do not approximate so nearly at their base, and the pterygoid fossæ are very shallow, whereas in *R. cuniculoides* they are deep. In the skull of the animal just mentioned there are two distinct longitudinal grooves on the palate, which extend backwards from the incisive foramina, and terminate in two rather large and deep excavations : these excavations are in the palatine bone, and situated between the last molar teeth ; they are separated from each other by a narrow, longitudinal, elevated ridge ; a narrow ridge also separates them from the pterygoid fossæ. At the bottom of each of these hollows are several minute foramina, and in front of them there are two larger longitudinal foramina. In *R. chinchilloides*, the longitudinal grooves on the palate and the posterior hollows are shallow, and consequently much less distinct ; the pterygoid fossæ are very nearly on the same plane as the palate, and are indicated only by a very slight depression. The incisor teeth are broader than in *R. chinchilloides*, and the molar teeth are proportionately smaller. The thin plate which forms the anterior root of the zygomatic arch is deeply emarginated in front in *R. cuniculoides* (see Plate 34, fig. 21, *b*.) ; but in *R. chinchilloides*, the anterior margin of this plate is nearly straight, (see Plate 34, fig. 20, *c*.)

In the form of the lower jaw of the two animals under consideration there are differences which will be more clearly understood upon comparing the figures. I will therefore merely notice one remarkable character which is found in *R. cuniculoides*, and that is, that the condyloid process is rather deeply concave on the inner side, a character which does not exist in *R. chinchilloides*, nor do I recollect having observed it in any other Rodent.

The principal dimensions of the skull of *R. chinchilloides*, are as follows :—

	In.	Lines.
Total length	1	2
Width	0	8½
Length of nasal bones	0	6⅛
of incisive foramina	0	4
Distance between the outer surface of the incisors and the first molar tooth, upper jaw	0	4½
Longitudinal extent of the three molars of the upper jaw, taken together	0	2¾
Length of a ramus of the lower jaw without the incisor	0	8

L

General Observations upon the foregoing Species of Muridæ.

In the foregoing descriptions I have endeavoured to convey an idea of the characters of the species of mice submitted to me for examination and description, by Mr. Darwin : there are, however, some points upon which I have been silent in my descriptions. I allude to the characters observable in the dentition. I have omitted to notice the various modifications in the structure of the molar teeth, because I found it would lengthen the descriptions to no good purpose, inasmuch as of almost all the species I have made outlines of the molars, which will convey a more clear idea than any verbal description can do.

Upon an inspection of the Plates, it will be seen, that by far the greater portion of the teeth figured, may be referred to one particular type of form or pattern, and that this pattern does not agree with that observed in the molars of *Mus Rattus*, *M. decumanus*, or *M. musculus*, whilst these three species agree essentially with each other.

In the young Black Rat (*Mus Rattus*), before the teeth are worn, the two anterior molar teeth, on either side of the upper jaw, present three longitudinal rows of tubercles, a central series of larger tubercles, and on each side of these, a row of smaller ones. The front molar has three of the larger tubercles arranged along the middle of the tooth ; three smaller ones on the outer side, and two, on the inner side. The second molars have two central tubercles, two outer, and two inner ones. The posterior molar is nearly round, the body of the tooth consists of three principal tubercles, and one small tubercle, situated on the inner and anterior portion of the tooth.

The corresponding teeth in the young of *Mus bimaculatus* present a very different appearance ; the molars, instead of having three longitudinal rows of tubercles, have only two. An idea of the appearance of these teeth may be formed by removing the inner row of tubercles from the molars of *Mus rattus*. We should then have, as in *Mus bimaculatus*, molars of a narrower form, the first tooth presenting six tubercles, the second, four ; and the posterior tooth devoid of the small inner lobe ; the opposing tubercles of each tooth, however, in *M. bimiculatus*, are of equal size.

The molars of the lower jaw of *Mus bimaculatus* agree with those of *M. Rattus* as to the number of tubercles which they possess ; they are, however, proportionately longer and narrower, and, when a little worn, these teeth, as well as those of the upper jaw, differ considerably from those of *M. Rattus*. In the last named animal, when the molars are slightly worn, the ridges of enamel run completely across the tooth, as in Figs. 18 and 19, Plate 34. Such is not the case in *M. bimaculatus* at any age. As soon as the molar teeth are worn, the folds of enamel penetrate the body of the tooth on each side, and those of one side alternate with those of the other,—in fact, they very nearly resemble those of the *Hamsters (Cricetus)*.

I have selected the molar teeth of *Mus Rattus* and *M. bimaculatus* for comparison, since I happened to possess specimens displaying both the young and adult states of each. But had I selected, on the one hand, almost any of the species brought from South America by Mr. Darwin, and, on the other hand, the *Mus musculus* or *M. decumanus*, I should have had to point out the same distinctions—the former agreeing in dentition with *M. bimaculatus*, and the latter with *M. Rattus*.

The differences pointed out, between the molar teeth of *Mus Rattus* and those of *M. bimaculatus*, I cannot but consider as important, since all the Old World species of *Mus* which I have yet had an opportunity of examining (and they are numerous) agree essentially with the former, whilst the only *Mus* from S. America (excepting *M. Musculus* and *M. decumanus*, which are carried in ships to all parts of the world) in which I have as yet found molar teeth like those of *M. Rattus*, is the *Mus Maurus*, and this it has been stated is possibly a variety of *M. decumanus*.

Although as yet I have not met with species in the Old World possessing the characters of the South American *Muridæ*, among those of North America, several have come under my observation. The *Mus leucopus, Symidon hispidum*, and the species of *Neotoma* certainly belong to the same group,* as does also the species of the Galapago Islands, described in this work under the name *Galapagoensis*.

These considerations have induced me to separate the South American mice from those of the Old World,—or rather from that group of which *M. decumanus* may be regarded as the type,—and to place them, together with such North American species as agree with them in dentition, in a new genus bearing the name *Hesperomys.*†

Whether this group be confined to the Western hemisphere or not, I will not venture to say, but I think I may safely affirm that that portion of the globe is their chief metropolis.

The species of the genus *Hesperomys*, which depart most from the type—whose dentition is least like figs. 5, *a*, and 5, *b*, Plate 33. or 6, *a*, and 6, *b*, of the

---

* I am acquainted with seven North American Species of *Muridæ*, all of which possess the dentition of *Hesperomys*.

† 'Εσπερος, West, and Μυς.

same Plate—recede still farther from the genus *Mus*, and approach more nearly (as regards the dentition) to the *Arvicolidæ*. Among the species here described I may mention as examples, *M. griseo-flavus*, *M. zanthopygus*, and *M. Darwinii*; —see the molar teeth figured in Plate 34. figs. 15, 16, and 17, —and among the North American species, those constituting the genus *Neotoma*. The latter make by far the nearest approach to the *Arvicolidæ* of any which have yet come under my observation, not only in the dentition, but in the form of the skull and the large size of the coronoid process of the lower jaw; there is, nevertheless, a tolerably well marked line of distinction between the crania of the *Arvicolidæ* and *Neotoma*.

The skulls of the animals belonging to the genera *Castor*, *Ondatra*, *Arvicola*, *Spalax*, and *Geomys*, which constitute the principal groups of the family *Arvicolidæ*, when compared with those of the family *Muridæ*, present, among others, the following distinctive characters.

The temporal *fossæ* are always much contracted posteriorly, by the great anterior and lateral development of the temporal bones; the plane of the inter-molar portion of the palate is below the level of the anterior portion; the coronoid process of the lower jaw is very large, the articular portion of the condyloid process is proportionately broad; the descending ramus, or posterior coronoid process, is so situated that its upper portion terminates considerably above the level of the crowns of the molars; this same process is generally * directed outwards from the plane of the horizontal ramus. The incisor teeth of the *Arvicolidæ* differ from those of the *Muridæ* in being proportionately broader and less deep from front to back—they are not laterally compressed as in *Mus*. The molar teeth are rootless,† and the folds of enamel are the same throughout the whole length of the tooth; whereas in *Mus* they enter less and less deeply into the body of the tooth as we recede from the crown, and towards the base of the visible portion (the tooth being in its socket) the indentations of the enamel are obliterated.

Now in the species of *Hesperomys*, the molar teeth are always rooted, and in the form of the skull and the lower jaw they agree with the *Muridæ*, and do not

---

* I am acquainted with only one exception, and that is in the genus *Castor*. In the genus *Ondatra*, the descending ramus is but slightly twisted outwards, but in all the other *Arvicolidæ*, whose crania I have examined, it is remarkably so, and in the genera *Spalax* and *Geomys*, where this character is carried to the extreme, the descending ramus projects from the alveolus of the long inferior incisors, in the form of a rounded and almost horizontal plate.

† In aged individuals of some of the species of *Arvicolidæ*, the molar teeth possess short roots. In a skull of *Ondatra* now before me I find all the molars divided at the base into two portions, which in all proba-bility would have formed solid roots had the animal lived longer.

present the characters above pointed out as distinguishing the *Arvicolidæ*, and as regards the cranium and lower jaw, it is only in the genus *Neotoma* that any approach is evinced.

Of the various groups of the order *Rodentia* found in South America, the *Sciuridæ*, so far as I am aware, are chiefly confined to the more northern parts, and do not occur in the most southern; the *Myoxidæ*, *Gerboidæ*, and *Arvicolidæ* are wanting. The species of the family *Muridæ* belong to different sections to those of the Old World. Of the *Leporidæ* I am acquainted only with one well established species—the *Lepus Braziliensis*, which however is not found "in tota America Australi," as Fischer says, there being no Hare yet found in the more southern parts, where the *Cavies* and *Chinchillas* appear to take their place. The remaining South American Rodents—certain species of *Hystricidæ*, the genera, *Echimys*, *Dasyprocta*, *Cælogenys* and *Myopotamus*, together with the *Octodontidæ* and *Chinchillidæ*, all possess a peculiar form of skull and of the lower jaw, (more or less approaching to figs. 1, Plate 33, and figs. 23, Plate 34.) which I have described in the "Magazine of Natural History," for February 1839, and which is rarely found in the North American, or Old World Rodents. In enumerating the above groups, I omitted the *Caviidæ*, because in the form of the lower jaw they differ somewhat from the rest—they possess, in fact, a form of lower-jaw peculiar to themselves; but in the Chinchillas[*] the transitions between one form and the other are found.

The South American *Muridæ*, which form the chief part of Mr. Darwin's collection, were none of them procured further north than latitude 30°, with the exception of those from the Galapagos Archipelago. The species occur at the following localities.

WEST COAST OF SOUTH AMERICA.	EAST COAST OF SOUTH AMERICA.
GALAPAGOS ARCHIPELAGO.	MALDONADO.
Mus Jacobiæ.	Mus decumanus.
—— Galapagoensis.	—— maurus.
	—— Musculus.
COQUIMBO.	—— tumidus.
	—— nasutus.
Mus longipilis.	—— obscurus.
—— Renggeri.	—— arenicola.
—— Darwinii.	—— bimaculatus.
	—— flavescens.
	Reithrodon typicus.

* See Proceedings of the Zoological Society for April 9th, 1839, p. 61.

WEST COAST.	EAST COAST.
VALPARAISO.	BUENOS AYRES.
Mus Renggeri.	—— Mus *decumanus.*
—— *decumanus.*	BAHIA BLANCA.
	Mus Braziliensis.
	—— elegans.
CONCEPCION.	—— gracilipes.
Mus longicaudatus.	RIO NEGRO.
	Mus griseo-flavus.
	PORT DESIRE.
CHILOE AND CHONOS ARCHIPELAGO.	Mus canescens.
Mus brachiotis.	ST. JULIAN.
	Reithrodon cuniculoïdes.
	—— xanthopygus.
	Reithrodon cuniculoïdes.
	SANTA CRUZ.
	Mus canescens.
	—— micropus.
	—— xanthopygus.
	Reithrodon cuniculoïdes.
	FALKLAND ISLANDS.
	Mus *decumanus.*
	—— *Musculus.*

STRAITS OF MAGELLAN.
Mus xanthorhïnus.
—— Magellanicus.
Reithrodon chinchilloïdes.

SECTION—HYSTRICINA.

FAMILY— ——?

MYOPOTAMUS COYPUS.

Myopotamus Coypus, *Auct.*

" This animal, in Chile, is known by the name of " Coypu ;" at Buenos Ayres, where an extensive trade is carried on with their skins, they are improperly called ' nutrias,' or otters. In Paraguay, according to Azara, their Indian name is ' guiya.' On the east side of the continent they range from Lat. 24° (Azara) to the Rio Chupat in 43° 20′;—distance of 1160 miles. This latter river is 170 miles south of the Rio Negro, and the intervening space consists of level, extremely arid, and almost desert plains, with no water, or at most one or two small wells. As the Coypu is supposed never to leave the banks of the rivers, and being, from its web-feet and general form of body, badly adapted for travelling on land, its occurrence in this river is a case, like so many others in the geographical distribution of animals, of very difficult explanation. The same remark is indeed applicable, but with less force, to its existence in the Rio Negro. On the west coast, it is found from the valleys of central Chile (Lat. 33°) to 48° S., or perhaps even somewhat farther, but not in Tierra del Fuego. So that, on the Atlantic side of the continent, the plains of Patagonia check its range southward, as, on the Pacific side, the deserts of Chile do to the north. Its range, including both sides, is from 24° to 48°, or 1440 miles. In the Chonos Archipelago these animals, instead of inhabiting fresh water, live exclusively in the bays and channels which extend between the innumerable small islets of that group. They make their burrows within the forest, a little way above the rocky beaches. I believe it is far from being a common occurrence, that the same species of any animal should haunt indifferently fresh water, and that of the open sea. We shall see that the Capybara is sometimes found on the islands near the mouth of the Plata; but these cannot be considered as their habitual station in the same manner as the channels in the Chonos Archipelago are to the Coypu. The inhabitants of Chiloe, who sometimes visit this Archipelago for the purpose of fishing, state that these animals do not live solely on vegetable matter, as is the case with those inhabiting rivers, but that they sometimes eat shell-fish. The Coypu is said to be a bold animal, and to fight fiercely with the dogs employed in chasing it. Its flesh when cooked is white and good to eat. An old female procured (January) amongst these islands, weighed between ten and eleven pounds." D.

Family—OCTODONTIDÆ.

Ctenomys Braziliensis.

Ctenomys Braziliensis, *De Blainville*, Bulletin de la Société Philomatique, June 1836, p. 62.

Maldonado, La Plata, (*June.*)

"This animal is known by the name of Tucutuco. I have given an account of its habits in my journal, but I shall here repeat it for the sake of keeping

together my observations on the less known animals. The Tucutuco is exceedingly abundant in the neighbourhood of Maldonado, but it is difficult to be procured, and still more difficult to be seen, when at liberty. Azara,* who has given
an account of its habits, with which every thing I saw perfectly agrees, states that
he never was able to catch more than one, although they are so extremely common. The Tucutuco lives almost entirely under ground, and prefers a sandy soil
with a gentle inclination; but it sometimes frequents damp places, even on the
borders of lakes. The burrows are said not to be deep, but of great length.
They are seldom open; the earth being thrown up at the mouth into hillocks not
quite so large as those made by the mole. Considerable tracts of country are
completely undermined by these animals. They appear, to a certain degree, to
be gregarious; for the man who procured my specimens had caught six together,
and he said this was a common occurrence. They are nocturnal in their habits;
and their principal food is afforded by the roots of plants, which is the object of
their extensive and superficial burrows. In the stomach of one which I opened I
could only distinguish, amidst a yellowish green soft mass, a few vegetable fibres.
Azara states that they lay up magazines of food within their burrows.

" The Tucutuco is universally known by a very peculiar noise, which it makes
when beneath the ground. A person, the first time he hears it, is much surprised;
for it is not easy to tell whence it comes, nor is it possible to guess what kind of
creature utters it. The noise consists in a short, but not rough, nasal grunt,
which is repeated about four times in quick succession; the first grunt is not so
loud, but a little longer, and more distinct than the three following: the musical
time of the whole is constant, as often as it is uttered. The name Tucutuco is
given in imitation of the sound. In all times of the day, where this animal is
abundant, the noise may be heard, and sometimes directly beneath one's feet.
When kept in a room, the Tucutucos move both slowly and clumsily, which
appears owing to the outward action of their hind legs; and they are likewise
quite incapable of jumping even the smallest vertical height. Mr. Reid, who
dissected a specimen which I brought home in spirits, informs me that the
socket of the thigh-bone is not attached by a ligamentum teres; and this ex-
plains, in a satisfactory manner, the awkward movements of their hinder extremities. Their teeth are of a bright wax yellow, and are never covered by the
lips: they are not adapted to gnaw holes or cut wood. When eating any thing,
for instance biscuit, they rested on their hind legs and held the piece in their fore
paws; they appeared also to wish to drag it into some corner. They were very
stupid in making any attempt to escape; when angry or frightened, they uttered

* Azara's Voyages dans l'Amerique Meridionale, vol. i. p. 324.

the Tucutuco. Of those I kept alive, several, even the first day, were quite tame, not attempting to bite or to run away; others were a little wilder. The man who caught them asserted that very many are invariably found blind. A specimen which I preserved in spirits was in this state; Mr. Reid considers it to be the effect of inflammation in the nictitating membrane. When the animal was alive, I placed my finger within half an inch of its head, but not the slightest notice was taken of it: it made its way, however, about the room nearly as well as the others. Considering the subterranean habits of the Tucutuco, the blindness, though so frequent, cannot be a very serious evil; yet it appears strange that any animal should possess an organ constantly subject to injury. The mole, whose habits in nearly every respect, excepting in the kind of food, are so similar, has an extremely small and protected eye, which, although possessing a limited vision, at once seems adapted to its manner of life.

" Several species probably will be found to exist south of the Plata. At Bahia Blanca (Lat. 39°) an animal burrows under ground in the same manner as the C. Braziliensis, and its noise is of the same general character, but instead of being double and repeated twice at short intervals. it is single and is uttered either at equal intervals, or in an accelerating order. I was assured by the inhabitants that these animals are of various colours, and, therefore, I presume that the two kinds of noises proceeded from two species. However this may be, they are extraordinarily numerous: many square leagues of country between the Sierras Ventana and Guetru-heigue are so completely undermined by their burrows, that horses in passing over the plain, sink, almost every step, fetlock deep. At the Rio Negro (Lat. 41°) some closely allied (or same?) species utters a noise, which is repeated only twice, instead of three or four times as with the La Plata kind. The sound is, moreover, louder and more sonorous; and so closely resembles that made in cutting down a small tree with an axe, that I have occasionally remained in doubt for some time to which cause to attribute it. Where the plains of Patagonia are very gravelly (as at Port Desire and St. Julian) the Ctenomys, I believe, does not occur; but at Cape Negro, in the Strait of Magellan, where the soil is damper and more sandy, the whole plain is studded with the little hillocks, thrown up by this destructive animal. It occurs likewise south of the Strait, on the eastern side of Tierra del Fuego, where the land is level. Captain King brought home a specimen from the northern side of the Strait, which Mr. Bennett* has called C. Magellanicus: it is of a different colour from the C. Braziliensis. I unfortunately did not make any note regarding the noise of this southern species: but the circumstance of its existence rather corroborates my belief in there being several other kinds in the neighbourhood of the Rio

---

* Transactions of the Zoological Society, vol. ii. p. 84.

Negro and Bahia Blanca. Otherwise we must believe that the same animal utters different kinds of noises, in different districts ; a fact which I should feel much inclined to doubt.

"Azara* says that the Tucutuco may be ' found every where ; provided that the soil be pure sand, and the situation not subject to be overflowed. As these conditions are fulfilled only in certain spots, their warrens are far separated from each other, even sometimes more than twenty-five leagues, without it being possible to conceive how these animals have been able to pass from one place to another.' The difficulty, I think, is much overstated ; for, as I have said, the burrows of the Tucutuco are sometimes made in very damp places, near lakes ; so that they certainly might pass over almost any kind of country. But if the *C. Braziliensis* and *C. Magellanicus* be considered as one species, as some French authors are inclined to do, then the difficulty will be increased in a very remarkable manner, as we shall be obliged to transport the Tucutuco over wide plains of shingle, and across many great rivers, and an arm of the sea."—D.

POEPHAGOMYS ATER.

Poephagomys ater, *F. Cuvier*, Annales des Sciences Naturelles,
2d series, Zoologie, tom. 1. p. 321.   June, 1834.

Chile, (*September.*)

"This animal is generally scarce, but in certain districts, I believe, of an alpine character, it is abundant.   It excavates very extensive superficial burrows, no doubt, for the purpose of feeding on the roots of plants, as in the case of the *Ctenomys Braziliensis*, the habits of which have just been described.   Horses passing over districts frequented by these animals, sink fetlock deep through the turf.   I procured my specimen from Valparaiso, where the country-people called it ' Cururo.' "—D.

OCTODON CUMINGII.

Octodon Cumingii, *Bennett*, Proc. of Committee of Science and Correspondence
of the Zool. Soc. for 1832, p. 46.
————————— Transactions of the Zoological Society of London, vol. ii. p. 81.   Pl. 16.
Dendrobius Degus, *Meyen*.   Acta Academiæ, c. l. c. Naturæ Curiosorum, xvi. p. 610.
Pl. 44, 1833.

Valparaiso, Chile, (*October.*)

* Azara Voyage dans l'Amerique Meridionale, vol. i. p. 324.

These little animals are exceedingly numerous in the central parts of Chile. They frequent by hundreds the hedge-rows and thickets, where they make burrows close together, leading one into another. They feed by day in a fearless manner; and are very destructive to fields of young corn; when disturbed, they all run together towards their burrows in the same manner that rabbits in England do when feeding outside a covert. When running they carry their tails high up, more like squirrels than rats; and they often remain seated on their haunches, like the former animals. According to Molina* they lay up a store of food for the winter, but do not become dormant. The Octodon is the "degu" of that author: he says that the Indians in past times used to eat them with much relish. These animals appear to be very subject to be piebald and albinos; as if partly under the influence of domestication.

Genus—ABROCOMA.†

Dentes primores ⅔ *acuti, eradicati, antice læves: molares utrinque* ¼ *subæquales, illis maxillæ superioris in areas duas transversales ob plicas vitreas acutè indentatus divisis; plicis utriusque lateris vix æquè profundis; illis mandibulæ inferioris in tres partes divisis, plicis vitreis his internè, semel externè indentatis, areá primá sagittæ cuspidem fingente, cæteris acutè triungularibus.*
Artus subæquales.
Antipedes 4-*dactyli, externo brevissimo, intermediis longissimis et ferè æqualibus.*
Pedes postici 5-*dactyli; digito interno brevissimo. Ungues breves et debiles, illo digiti secundi lato et lamellari; omnibus setis rigidis obtectis.*
Caput mediocre, auribus magnis, membranaceis; oculis mediocribus.
Cauda breviuscula.
Vellus perlongum, et molle.

The genus *Abrocoma* is evidently allied on the one hand to the genera *Octodon, Poephagomys,* and *Ctenomys,* and on the other to the family *Chinchillidæ.* The four genera just mentioned possess so many characters in common, that it would be well to unite them, and the name *Octodontidæ* may be used to designate the group.

The *Octodontidæ* appear to bear the same relations to *Echimys,* as the *Arvicolæ* do to the *Muridæ.*

* Compendio de la Hist. Nat. del Reyno de Chile, vol. i. p. 343.
† 'Αβρος, soft; Κομη, hair.

In the *Octodontidæ* the skull is rather short, the inter-orbital space is broad ; the ant-orbital passage is large ; the zygomatic arch is thrown out horizontally from the plane of the palate ; the malar bone is broad and somewhat compressed, and throws up a small post-orbital process ; the glenoid cavity of the temporal bone is narrow ; the palate is contracted, and deeply notched posteriorly, the portion which lies between the molar teeth descends below the level of the anterior portion ; the incisive foramina are wide : the body of the anterior and posterior sphenoids is very narrow; and the foramina on either side of them are large : the occipital condyles are very narrow, widely separated, and the articular surface is nearly vertical.* The descending *ramus* of the lower jaw springs from the outer side of the alveolar portion, and terminates in a point, more or less acute.

The incisors of the upper and lower jaws are of the same width : the molars are $\frac{4-4}{4-4}$, rootless.

In external characters the species of the present group vary considerably. The toes are 5|5 or 4|5. The claws of the hind feet are covered by strong, curved bristly hairs.

The principal points of distinction in the external characters of the four genera under consideration, may be thus expressed.

† TOES 5|5.

A. Fore feet formed for burrowing—strong and armed with large claws; tail short.

    a. Ears minute, incisors very broad .   .   .   .   .   .   *Ctenomys.*

    b. Ears small, incisors broad .   .   .   .   *Poephagomys.*

B. Fore feet weak ; claws small ; incisors narrow ; ears large.

    a. Tail with the apical portion furnished with long hair . ,   .   .   *Octodon.*

†† TOES 4|5.

    b. Tail furnished throughout with short adpressed hairs .   .   .   .   *Abrocoma.*

It is not only in the comparatively small size and weakness of the fore feet that *Abrocoma* approaches more nearly to *Octodon ;* but it agrees, in having the soles, both of the fore and hind feet (which are devoid of hair), covered with minute round fleshy tubercles (see the under side of the tarsus figured in Plate 28.)

In *Octodon,* however, the toes have on their under side transverse incisions, as the *Muridæ,* and many other Rodents ; a character not found in *Abrocoma.*

---

* There is a wide difference between the present animals and the *Arvicolidæ* in the form of the occipital condyles : the same difference is also observable between *Echimys* and *Mus.* The *Octodontidæ* in fact have the same form of condyles as the Chinchillas and Cavies. In this and many other characters the last mentioned animals evince an affinity to the *Leporidæ.*

Abrocoma Bennettii

Here the under-side of the toes, like the sole of the foot, is covered with minute tubercles.

Though in the form of the skull *Abrocoma Cuvieri*[*] agrees most nearly with that of *Octodon ;* it differs in having the anterior portion narrower and rather larger, compared to the part devoted to the protection of the brain ; the zygomatic arch is shorter, the incisive foramina are longer, the body of the anterior sphenoid is narrower, and the auditory bullæ are larger. The principal differences observable in the form of the lower jaw of *Abrocoma*, when compared with that of *Octodon*, consists in the coronoid process being smaller, the condyloid narrower from front to back ; the descending *ramus* more deeply emarginated posteriorly, and the angle longer and more attenuated.

In those characters in which the skull of *Abrocoma* departs from that of *Octodon*, it approaches nearer to *Chinchilla*. In the peculiar form and large size of the ears, in the extreme softness of the fur, in the greater development of the pads on the under side of the toes, and in the possession of only four toes to the fore feet, there are other points of resemblance between *Abrocoma* and *Chinchilla*. In the Chinchilla as well as in *Octodon* and *Abrocoma*, we find the toe corresponding to the second (counting from the inner side) furnished with a broad hollow nail ;[†] there are also stiff bristly hairs covering this nail as in the *Octodontidæ*.

The extreme softness of the fur of the animals about to be described, suggested for them the generic name of *Abrocoma*. The fur consists of hairs of two lengths, and the longer hairs are so extremely slender that they might almost be compared to the web of the spider. The specific names applied are those of the distinguished naturalists who first made us acquainted with the two genera, *Octodon* and *Poephagomys*.

1. ABROCOMA BENNETTII.

PLATE XVIII.

Abrocoma Bennettii, *Waterh.*, Proceedings of the Zoological Society of London, for February 1837, p. 31.

A. corpore suprà griseo, ad latera pallidiore et pallidè cervino lavato, subtùs albescenti-cervino ; gulá albescenti-griseá ; pedibus sordidè albis: auribus amplis, ad marginem posticum rectis, extùs ad bases vellere, sicùt in corpore, obsitis: caudá corpore breviore, ad basin crassiusculá, pilis brevibus incumbentibus vestitá.

DESCRIPTION.—Form stout ; ears large, with the posterior margin straight ; fore

* I have not had an opportunity of examining the skull of *Abrocoma Bennettii*.
† This nail no doubt is used to cleanse the fur, and the bristly hairs may also assist in the operation; the two small toes of the Kangaroo's hind foot are used for the same purpose.

feet rather small, tarsus short ; tail rather shorter than the body, thick at the base; fur long and extremely soft, and silk-like. General colour pale grayish brown, with a slight yellow wash ; the upper part of the head and the back dusky brown; under parts of the body very pale yellowish brown, inclining to white ; chin and throat whitish ; feet dirty white ; tail well clothed with hairs, which are closely adpressed, brown above, and of a very pale brown beneath at the base, darker towards the apex. The hairs of the moustaches are numerous, long, rather slender, and of a brownish colour. The ears are brown, furnished externally at the base with fur resembling that of the body ; the remaining parts (both external and internal) are beset with long and extremely slender brown hairs, which project considerably beyond the margin of the ear. The ordinary fur on the back is about ten lines in length, but thickly interspersed with this fur, are longer hairs which are so delicate that they may almost be compared to the spiders' thread. Both on the upper and under side of the body the fur is deep gray at the base. The incisors are yellow.

	In.	Lines.			In.	Lines.
Length from nose to root of tail	. 9	9	Length of tarsus (claws included) .	. 1	4
of tail	. 5	0	of ear	. 0	10
from nose to ear . .	. 1	11	Width of ear	. 1	0½

Habitat, Chile, (*August.*)

" This animal was caught amongst some thickets in a valley on the flanks of the Cordillera, near Aconcagua. On the elevated plain, near the town of Santa Rosa, in front of the same part of the Andes, I saw two others, which were crawling up an acacia tree, with so much facility, that this practice must be, I should think, habitual with them."—D.

2. ABROCOMA CUVIERI.

PLATE XXIX.

Abrocoma Cuvieri, *Waterh.*, Proceedings of the Zoological Society of London for February 1837, p. 32.

Ab. suprà grisea, levitèr ochraceo lavata; abdomine gulâque albescenti-griseis; pedibus sordidè albis; auribus amplis, ad marginem posticum distinctè emarginatis; caudâ corpore multò breviore, et nigrescente.

DESCRIPTION.—Ears large ; tail considerably shorter than the body ; fur extremely

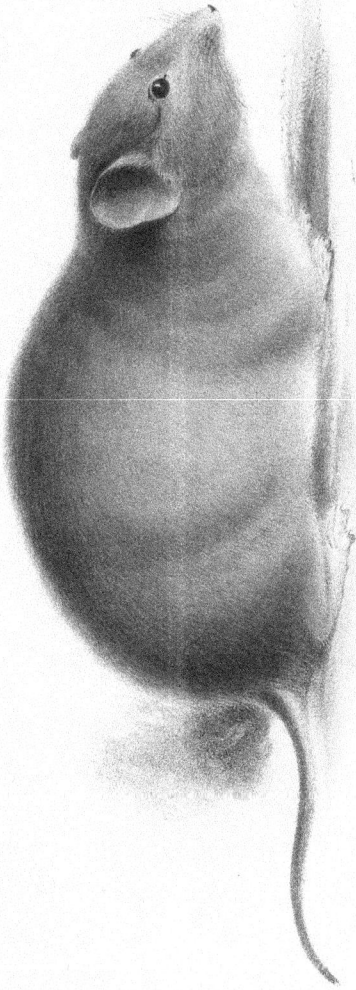

Myocoma Cavia

soft; general colour gray faintly washed with yellow; under parts of the body grayish white; feet dirty white; tail dusky, paler beneath at the base: the ears are large, distinctly emarginated behind, and appear to be almost naked, but, upon close examination, long and extremely fine hairs may be observed. All the fur is gray at the base; the hairs of the moustaches are numerous and very long, those nearest the mouth are white, the others are black at the base and grayish beyond. The incisors are of a palish yellow colour.

	In.	Lines.			In.	Lines.
Length from nose to the root of tail	6	6	Length of tarsi (claws included)	.	1	1
of tail	2	10	of ear	.	0	7
from nose to ear	1	4	Width of ear	.	0	7½

Habitat, Chile, (*September.*)

This species is about one-third the size of the last, it differs moreover in being gray instead of brown, and in having the posterior margin of the ear emarginated; the tail is also rather shorter in proportion.

The skull* is figured in Plate 33, fig. 1, *a*, and 1, *b*; and fig. 23, *a*, Plate 34. Its length is 1 inch, 4½ lines; width 9¼ lines; length of nasal bones 6 lines; distance between fore part of incisors and the front molar (upper jaw) 5 lines; longitudinal extent of the three molars of upper jaw 3 lines; length of auditory bullæ 5¾ lines; length of *ramus* of lower jaw (see Plate 33, fig. 1, *c*,), without incisors, 11½ lines. Fig. 23, *c*, Plate 34, represents the inner side of a *ramus* of the lower jaw: fig. 1, *d*, Plate 33, is the lower jaw seen from above: fig. 23, *b*, Plate 34, is the same seen from beneath. This view is given to show the position of the descending ramus of the lower jaw—that it springs from the outer side of the alveolar portion, as in a great portion of the South American Rodents, such as *Dasyprocta*, *Myopotamus*, *Echimys*, *Chinchilla*, and also in that genus found in the West Indian islands, *Capromys*. Fig. 1, *e*, Plate 33, represents the molar teeth of the upper jaw, and fig. 1, *f*, those of the lower.

" This species is abundant on the dry hills, partly covered with bushes, near Valparaiso."—D.

* The skull is, unfortunately, imperfect, the hinder portion is injured, and the arches which enclosed the ant-orbital openings are broken.

Family—CHINCHILLIDÆ.

Lagostomus trichodactylus.

Lagostomus trichodactylus, *Brooks*, Transactions of the Linnean Society, vol. xvi. p. 95, Pl. 9.

La Vizcacha, *Azara*, Essais sur l'Histoire Naturelle des Quadrupedes de la Province du Paraguay, vol. ii. p. 41, Trad. Franc.

Vischacha, *Meyen*, Acta Academiæ, c. l. c. Naturæ Curiosorum, Tom. xvi. pars 2, p. 584.

Habitat, La Plata.

" I will not repeat what I have said about the habits of this animal in my Journal, as it is merely a corroboration of Azara's account. According to that author, the Bizcacha is not found north of 30°; and its southern limit occurs in the neighbourhood of the Rio Negro in 41°. Where the plains are gravelly, it is not abundant, but (differently from the *Cavia Patagonica*,) it prefers an argillaceous and sandy formation, such as that near Buenos Ayres. The Bizcacha abounds over the whole Pampas, even to the neighbourhood of Mendoza, and there it is replaced in the Cordillera by an Alpine species. Of the latter animal, I saw one seated on a pinnacle at a great height, but I could not obtain a specimen of it. Azara* has remarked that the Bizcacha, fortunately for the inhabitants of Banda Oriental, is not found to the eastward of the Rio Uruguay; and what makes the case more remarkable is, that although thus bounded by one river, it has crossed the broader barrier of the Parana, and is numerous in the province of Entre Rios. I was assured by a man, whose veracity I can perfectly trust, that these animals, quasi canes, post coitum adnexi sunt."—D.

Family—CAVIIDÆ.

Kerodon Kingii.

Kerodon Kingii, *Bennett*, Proceedings of the Zoological Society of London for 1835, p. 190.

Habitat, Patagonia.

" The Kerodon is common at intervals along the coast of Patagonia, from the

* Azara ' Voyages dans l'Amerique Meridionale,' vol. i. p. 316.

Rio Negro (Lat. 41°) to the Strait of Magellan. It is very tame, and commonly feeds by day: it is said to bring forth two young ones at a birth. At the Rio Negro it frequents in great numbers the bottoms of old hedges: at Port Desire it lives beneath the ruins of the old Spanish buildings. One old male killed there weighed 3530 grains. At the Strait of Magellan, I have seen amongst the Patagonian Indians, cloaks for small children made with the skins of this little animal; and the Jesuit Falkner says, that the people of one of the southern tribes, take their name from the number of these animals which inhabit their country. The Spaniards and half-civilized Indians, call the Kerodon, 'conejos,' or rabbit; and thus the mistake has arisen, that rabbits are found in the neighbourhood of the Strait of Magellan."—D.

1. Cavia Cobaia.

Cavia Cobaia, *Auct.*

Habitat, Maldonado, La Plata, (*June.*)

" This animal, known by the name of Aperea, is exceedingly common in the neighbourhood of the several towns which stand on the banks of the Rio Plata. It frequents different kinds of stations,—such as hedge-rows made of the Agave and Opuntia, or sand-hillocks, or again, marshy places covered with aquatic plants;—the latter appearing to be its favourite haunt. Where the soil is dry, it makes a burrow; but where otherwise, it lives concealed amidst the herbage. These animals generally come out to feed in the evening, and are then tame; but if the day be gloomy, they make their appearance in the morning. They are said to be very injurious to young trees. An old male killed at Maldonado, weighed 1 lb. 3 oz. In all the specimens I saw there, (during June, or winter,) I observed, that the hair was attached to the skin less firmly than in any other animal I remember to have seen."—D.

2. Cavia Patachonica.

Cavia Patachonica, *Shaw*, General Zoology, vol. ii., part 1, p. 226.
Dasyprocta Patachonica, *Desmarest*, Mamm. p. 358, Sp. 574.
Dolichotis——————————————— in Note, p. 359-360
Chloromys Patachonicus, *Lesson*, Manuel de Mammalogie, p. 301.
Lièvre Pampa, *Azara*, Essais sur l'Histoire Naturelle des Quad. de la Province
du Paraguay. French Translation, vol. ii. p. 51.

In the form of the cranium, and in the structure of the teeth, this animal possesses all the characters of the Cavies (*Caviidæ*).*

Habitat, Patagonia.

* See Proceedings of the Zoological Society for April, 1839, p. 61.

N

"This animal is found only where the country has rather a desert character. It is a common feature in the landscape of Patagonia, to see in the distance two or three of these Cavies hopping one after another in a straight line over the gravelly plains, thinly clothed by a few thorny bushes and a withered herbage. Near the coast of the Atlantic, the northern limit of this species is formed by the Sierra Tapalguen, in latitude 37° 30', where the plains rather suddenly become greener and more humid. The limit certainly depends on this change, since near Mendoza, (33° 30'.) four degrees further northward, where the country is very sterile, this animal again occurs. Azara erroneously supposed that its northern range was only 35°.* It is not clear on what circumstances its limit southward between Ports Desire and St. Julian (about 48° 30'.) depends ; for there is in that part no change in the features of the country. It is, moreover, a singular circumstance, that although the Cavy was not seen at Port St. Julian during our voyage, yet Capt. Wood, in 1670, speaks of them as being numerous there. What cause can have altered, in a wide, uninhabited, and rarely visited country, the range of an animal like this ?

"Azara states,† that the Cavy never excavates its own burrow, but uses that of the Bizcacha. Wherever this animal is present, without doubt this is true ; but on the sandy plains of Bahia Blanca, where the Bizcacha is not found, the Spaniards maintain that the Cavy is its own workman. The same thing occurs with the little owls of the Pampas (*Noctua cunicularia*), which have been described by travellers as standing like sentinels at the mouths of almost every burrow ; for in Banda Oriental, owing to the absence of the Bizcacha, these birds are obliged to hollow out their own habitations. Azara says, also, that this Cavy, except when pressed by danger, does not enter its burrow ; on this point I must again differ from that high authority. At Bahia Blanca I have repeatedly seen two or three of these animals sitting on their haunches by the mouths of their holes, which they quietly entered as I passed by at a distance. Daily, in the neighbourhood of these spots, the Cavies were abundant : but differently from most burrowing animals, they wander, commonly two or three together, to miles or leagues from their home ; nor do I know whether they return at night. The Cavy feeds and roams about by day ; is shy and watchful ; seldom squats after the manner of a hare ; cannot run very fast, and, therefore, is frequently caught by a couple of dogs, even of mixed breed. Its manner of running more resembles that of a rabbit than of a hare. The Cavy generally produces two young ones at a birth, which are brought forth within the burrow. The flesh, when cooked, is

---

* Azara, Voyage dans l' Amérique Méridionale, vol. i. p. 318.
† Azara, Quadrupeds of Paraguay.

very white; it is, however, rather tasteless and dry. Full grown animals weigh between twenty and twenty-six pounds. '—D.

HYDROCHŒRUS CAPYBARA.

Hydrochœrus Capybara, *Auct.*

"These animals are common wherever there are large rivers or lakes, over that part of the South American Continent which lies between the Orinoco and the Plata, a distance of nearly 1400 miles. They are not generally supposed to extend south of the Plata; but as there is a Laguna Carpincho (the latter being the provincial name of the Capybara) high up the Salado, I presume they have sometimes been seen there. Azara does not believe they ever frequent salt water; but I shot one in the Bay of Monte Video; and several were seen by the officers of the Beagle on the Island of Guritti, off Maldonado, where the water is very nearly as salt as in the sea. The one I shot, at Monte Video, was an old female; it measured from tip of snout to end of stump-like tail, 3 feet 8½ inches, and in girth 3 feet 2 inches. She weighed 98 pounds. I opened the stomachs of a couple, which I killed near a lake at Maldonado, and found them distended with a thin yellowish-green fluid, in which not more than a trace of a vegetable fibre could be distinguished : it is in accordance with this fact, that a part of the æsophagus is so narrow, as I am informed by Mr. Owen, that scarcely anything larger than a crow-quill can be passed down it. The shape of the dung of these animals is a short straight cylinder, rounded at the extremities ; when dried and burnt, it affords a pleasant smell like that from cedar wood. These animals do not burrow holes, but live amongst the thickets, or beds of rushes near rivers and lakes. At Maldonado they often may be seen during the day, seated on the grassy plain in small groups of three and four, at the distance of a few yards from the border of the lake, which they frequent. I must refer the reader for a few more details respecting their habits, to my Journal of Researches.—D."

Section — LEPORINA.

Family—LEPORIDÆ.

Lepus Magellanicus.

Lepus Magellanicus, *Lesson et Garnot*, Zoologie du Voyage autour du Monde de la Corvette, La Coquille.

" A black variety of the domesticated species, which was turned out on these islands by the earlier colonists, has been considered, but with some hesitation, by M. Lesson, as a distinct species. He has called it *Lepus Magellanicus*, and has given the following specific character, —' *Pilis omnino atro-violaceis, albis passim sparsis: auriculis fuscis, capite brevioribus; maculâ albâ naso, interstitio narium, menti, gulæ, frontique.*'* In the specimens preserved on board the Beagle, the form and position of the white marks neither agree with M. Lesson's description, nor with each other. In one there is a broad white patch on one side of the head, and another on one of the hinder thighs. The Spaniards employed in hunting wild cattle, (who are all excellent practical observers) assured me, that the black rabbits were only varieties of the common gray kind, and they gave the following reasons for thinking so;—namely, that the two sorts did not live apart; that the black one had not a different range from the other; that the two bred freely together, and that they produced pie-bald offspring. As the rabbits extend their range very slowly, (not having yet crossed the central range,) the Spaniards have sometimes carried a few aud turned them out in different parts of the island, and thus they have ascertained that the black and gray kinds breed together freely. Bougainville, moreover, who visited the part of the island, where the black variety is now most common, distinctly states, in his voyage round the world, that no animal, excepting the great wolf-like fox inhabited these islands. M. Lesson supposes that the *Lepus Magellanicus* is found near the Strait of Magellan; but I inquired of the Indians, who live there, and they knew of no other 'conejos' or rabbits, except the *Kerodon Kingii*, which no doubt is the animal alluded to by the early voyagers."—D.

1. Dasypus hybridus.

Dasypus hybridus, *Auct.*

" This species seems to prefer rocky and slightly undulating ground, and

* Voyage de La Coquille. Partie Zoologique, vol. i. p. 168.

hence is common in Banda Oriental and Entre Rios. Azara says it is found from 26° 30', to at least 41° south ; but, I was assured, perhaps incorrectly, that the Sierra Tapalguen (37° 30'), where the nature of the country becomes slightly different, is its southern limit. The *D. villosus, minutus,* and *mataco,* are found at Bahia Blanca, in latitude 39°. I was also assured that these three species, together with the *D. hybridus,* frequent the plains near Mendoza, in latitude 33° to 34°."—D.

2. DASYPUS MINUTUS.

Dasypus minutus, *Auct.*

" The northern limit of this species on the Atlantic side of the continent, is (as I was told by the inhabitants) near the southern one of the *D. hybridus,* namely, 37° 30'. It is extremely abundant on the arid plains near the Sierra Ventana, and likewise in the neighbourhood of the Rio Negro. This species has a range considerably further southward than any other : I obtained specimens at Port Desire, where, however, it is far from common, and at Santa Cruz (in latitude 50°) I saw its tesselated covering lying on the ground. At Bahia Blanca, I found in the stomach of this armadillo, coleoptera, larvæ, roots of plants, and even a small snake of the genus Amphisbæna. All the species, excepting one, wander about by day. At Bahia Blanca, during a morning's ride, three or four of the *D. minutus* generally were met with ; but, in order to secure them, it was necessary to jump off one's horse as quickly as possible, otherwise, they would have disappeared by burrowing in the sandy soil. This species often endeavours to escape detection by squatting close to the ground, and remaining motionless."—D.

1. DIDELPHIS AZARÆ.

Didelphis Azaræ, *Auct.*

" This species is said to inhabit burrows : it is nocturnal, and is very destructive to poultry. The body after death possesses a very offensive odour. My specimen was procured at Maldonado."—D.

2. Didelphis crassicaudata.

Plate XXX.

Didelphis crassicaudata, *Desmarest*, Nouv. Dict. d'Hist. Nat. 2d Ed. ix. p. 425.
———— ———————— Mammalogie, p. 257, Species 393.
Microuré troisième, ou Macrouré à grosse queue, *Azara*, Essais sur l'Histoire Nat. des
　　　　Quad. de la Province de Paraguay, vol. i. p. 284.

*D. capite brevi ; auribus parvis ; colore corporis fuscescenti-flavo subtùs pallidiore ;
infra oculos flavescente ; caudâ crassâ, caput corpusque, quoad longitudinem, ferè
æquante ; ad basin corporis colore tinctâ, dein nigra, ad apicem albâ.*

Description.—Head short; ears small, the posterior edge emarginated near the
base, distinctly furnished with hairs; tail slightly exceeding the body in
length, very thick at the base; tarsi small; fur moderately long, slightly
harsh, and somewhat adpressed (much less woolly than in most Opossums):
general tint brownish yellow, under parts paler; anterior angle of the eye
and muzzle brown, the tip of the chin, and also the tip of the muzzle on either
side whitish; on the cheeks, a little below the eyes, is a patch of yellow which
extends round the angle of the mouth : about one-third of the tail is covered
with fur of the same colour and character as that on the body ; beyond this
the tail is black, excepting a small portion, about one inch in length, at
the apex, which is white ; and the hairs are short, closely adpressed,
and scarcely hide the scales which are beneath : the fore portion of each
foot is brown : the hairs covering the ears on the outer side are brownish,
and those on the inner side of the ear are yellow, but towards the outer
margin they are brown.　The hairs of the back have the basal half gray, and
the apical half ochreous, terminating in yellowish brown ; on the belly and
underside of neck, the hairs are ochreous, faintly tinted with gray at the
base.

	In.	Lines.				In.	Lines.
Length from nose to root of tail	. . 1	3		Length of tarsus	. . . . .	1	5½
of tail	 10	3		of ear	. . . .	0	6
from nose to ear	. . . 2	1½					

Habitat, Maldonado, La Plata, (*June*).

The species described by Azara, under the name *Macrouré à grosse queue*,
agrees so perfectly with the present animal, that I have no hesitation in referring

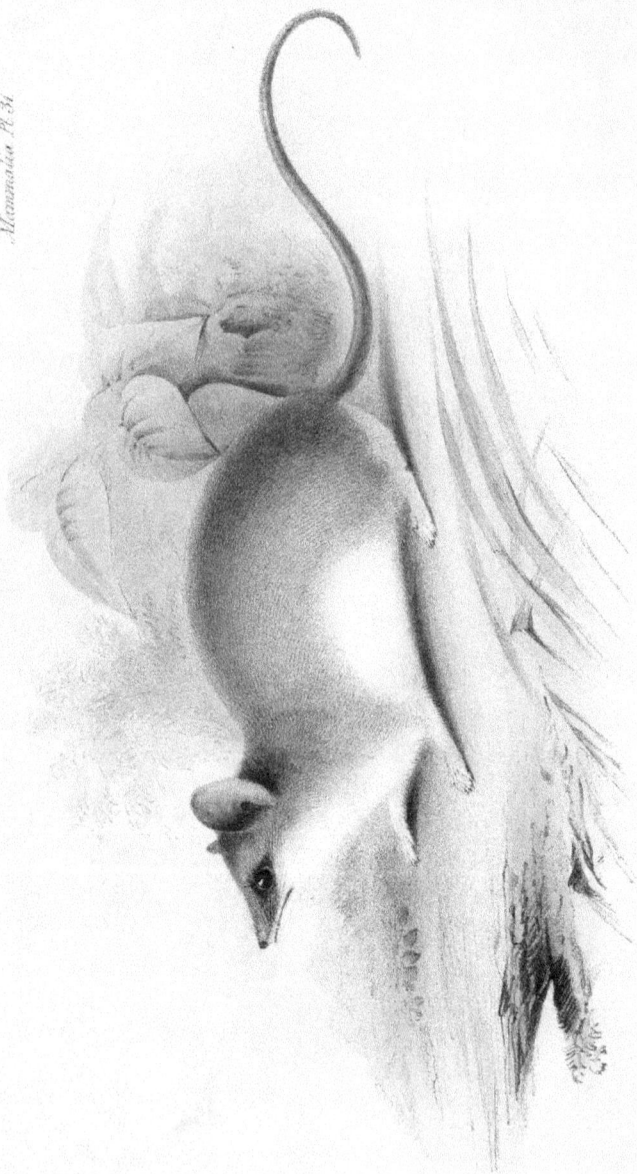

Didelphis elegans

it to the *Didelphis crassicaudata* of Desmarest, which is founded upon Azara's description.

The head of the *Didelphis crassicaudata* is shorter and less pointed than in most other Opossums; the ears are unusually small, and the tail is very thick. In the character of the fur also, this species differs from most others, the hairs being rather short and somewhat adpressed; and the soft under-fur being very scanty. Upon separating the fur on the back and sides of the body, numerous young hairs were visible in the specimen from which the above description is taken, and these were of a bright rusty red tint; the colouring of the animal therefore would, in all probability, have been very different after a short time, had it not been killed. Those observed by Azara varied considerably in their colouring. The skull is figured in Plate 34. figs. 25. Fig. *d* represents a *ramus* of the lower jaw.

	In.	Lines.
Length of the skull	2	4
Width	1	3
Length of nasal bones	0	9½
———— of palate	1	2¾
Width of palate between the posterior molars	0	5
Distance between forepart of front incisors and forepart of canine	2	0¾
Distance between forepart of canine and hinder part of last molar	1	0
Length of *ramus* of lower jaw (incisors not included) . .	1	10½

" This specimen was caught at Maldonado: it weighed 14½ oz."—D.

3. DIDELPHIS ELEGANS.

PLATE XXXI.

D. vellere longo et molli, corpore suprà cinereo-fuscescente lavato; pedibus corporeque subtùs albis, oculis nigro circumdatis, interspatio cinerescente; auribus magnis fuscescentibus; caudâ, capite et corpore, paulo breviore.

DESCRIPTION.—Muzzle slender and pointed; ears large; tail rather shorter than the head and body taken together; fur long and very soft: general tint of the upper parts of the head and body ashy-gray washed with brown; on the sides of the body, especially near the shoulders, a faint yellowish tint is observable; the lower part of the cheeks, the throat, under parts of the body and the feet, are white, with an indistinct yellowish tint; the eyes are encircled with brownish-black, which tint is extended forwards on to the sides of the muzzle; the upper surface of the muzzle and the inter-orbital space is

pale. The tail is furnished throughout with minute decumbent hairs, excepting
a small naked space at the tip beneath, of about one line in length ; on the
upper surface they are brown, and on the under, they are whitish. The
fur of the upper and under parts of the body is deep gray at the base ; on
the lower part of the cheeks, chin, and on the mesial line of the throat and
chest, the hairs are uniform—not gray at the base. The ears are brown,
and to the naked eye, appear naked.

	In.	Lines.		In.	Lines.
Length from nose to root of tail : .	4	6	Length from nose to ear . . .	1	1½
of tail	4	4	of ear	0	7¼
of tarsus (claws included) . .	0	7½	width of ear	0	7½

Habitat, Valparaiso, Chile, (*October.*)

This little Opossum, which is the only species I am acquainted with from the
west side of the Cordillera, was exhibited at one of the scientific meetings of the
Zoological Society, and its characters were pointed out by Mr. James Reid, who
proposed for it the specific name of *hortensis*,* a name which was given from the
circumstance that in Mr. Darwin's notes it is stated that a small Opossum was
found in a garden at Maldonado. These notes however refer to the *Didelphis
brachyura.* The skull of this animal is figured in Plate 35. Fig. 5, *a*, represents the upper side ; 5, *b*, the under side ; and 5, *c*, is the side view. Fig. 5, *d*, is
the lower jaw, and 5, *e*, is the same magnified. The length of the skull is 14¼ lines ;
width, 8 lines ; length of palate, 7¼ lines ; inter-orbital space, 2¼ lines ; length of
ramus of lower jaw, 10½ lines. In the palate are two long openings which commence
opposite the posterior false molar, and terminate opposite the hinder portion of the
penultimate true molar : the incisive foramina are nearly one line in length. On
the posterior portion of the palate there are four other foramina, one on each side
near the. posterior molar, and one on either side the mesial line, behind the large
palatine openings above mentioned.

"These little animals frequent the thickets growing on the rocky hills, near
Valparaiso. They are exceedingly numerous, and are easily caught in traps
baited either with cheese or meat. The tail appeared to be scarcely at all
used as a prehensile organ ; they are able to run up trees, with some degree of
facility. I could distinguish in their stomachs the larvæ of beetles."—D.

* See Proceedings of the Zoological Society of London for January, 1837, p. 4.; its characters were not
published.

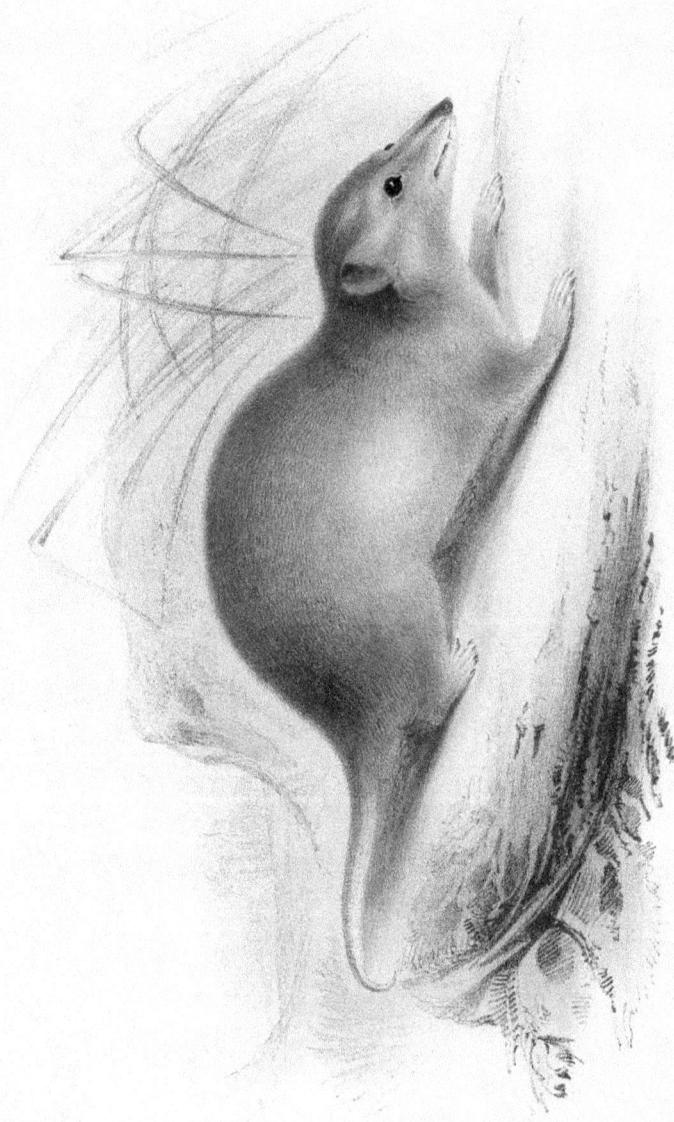

Didelphis brevicaudata

4. DIDELPHIS BRACHYURA.

PLATE XXII.

Didelphis brachyura, *Auct.*

D. vellere brevi, corporis suprà cinereo, flavo lavato ; lateribus capitis, corporisque, et partibus inferioribus rufescenti-flavis, gulâ et abdomine pallidioribus ; caudâ brevi.

DESCRIPTION.—Head large; canine teeth very large; ears rather small; tail short, rather more than half the length of the body; fur short and crisp ; the back and upper surface of the head ashy gray, grizzled with yellowish white ; the sides of the head and body, and under parts rusty yellow, rather paler on the belly than on other parts, and of a deeper hue on the rump and cheeks; the eye is encircled with rusty yellow ; feet yellowish ; tail clothed with short stiff hairs, and exhibiting scales, brownish above, and dirty yellowish white beneath—a small naked space beneath, at the tip, of about two lines in length. Fur of the back grayish at the base, that on the belly uniform ; ears clothed with minute yellowish white hairs.

	In.	Lines.		In.	Lines
Length from nose to the root of tail	6	0	Length of tarsus (claws included) . .	0	$8\frac{3}{4}$
from nose to ears . . .	1	6	of ear	0	$3\frac{3}{4}$
of tail	2	8			

Habitat, Maldonado, La Plata, (*June.*)

Never having seen a good figure of this animal, I have thought it desirable to introduce it in the plates of this work.

The *Didelphis brachyura* is closely allied to the *D. tricolor* of authors, but in that species the upper parts of the body are nearly black ; the sides of the head and body are of a deep rusty red tint, and the under parts are almost white.

" Was caught by some boys digging in a garden. Its intestines were full of the remains of insects, chiefly ants and others of the Hemipterous order."—D.

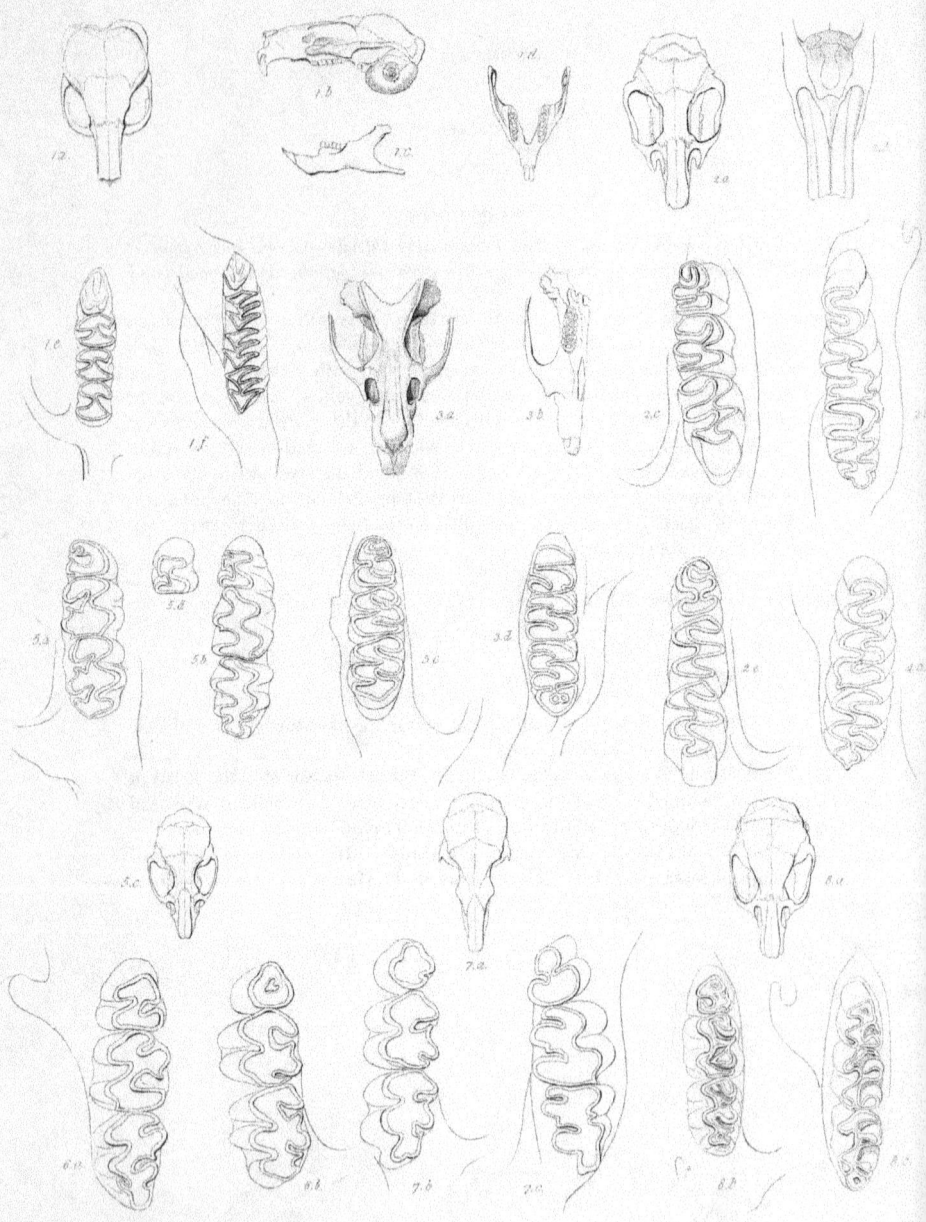

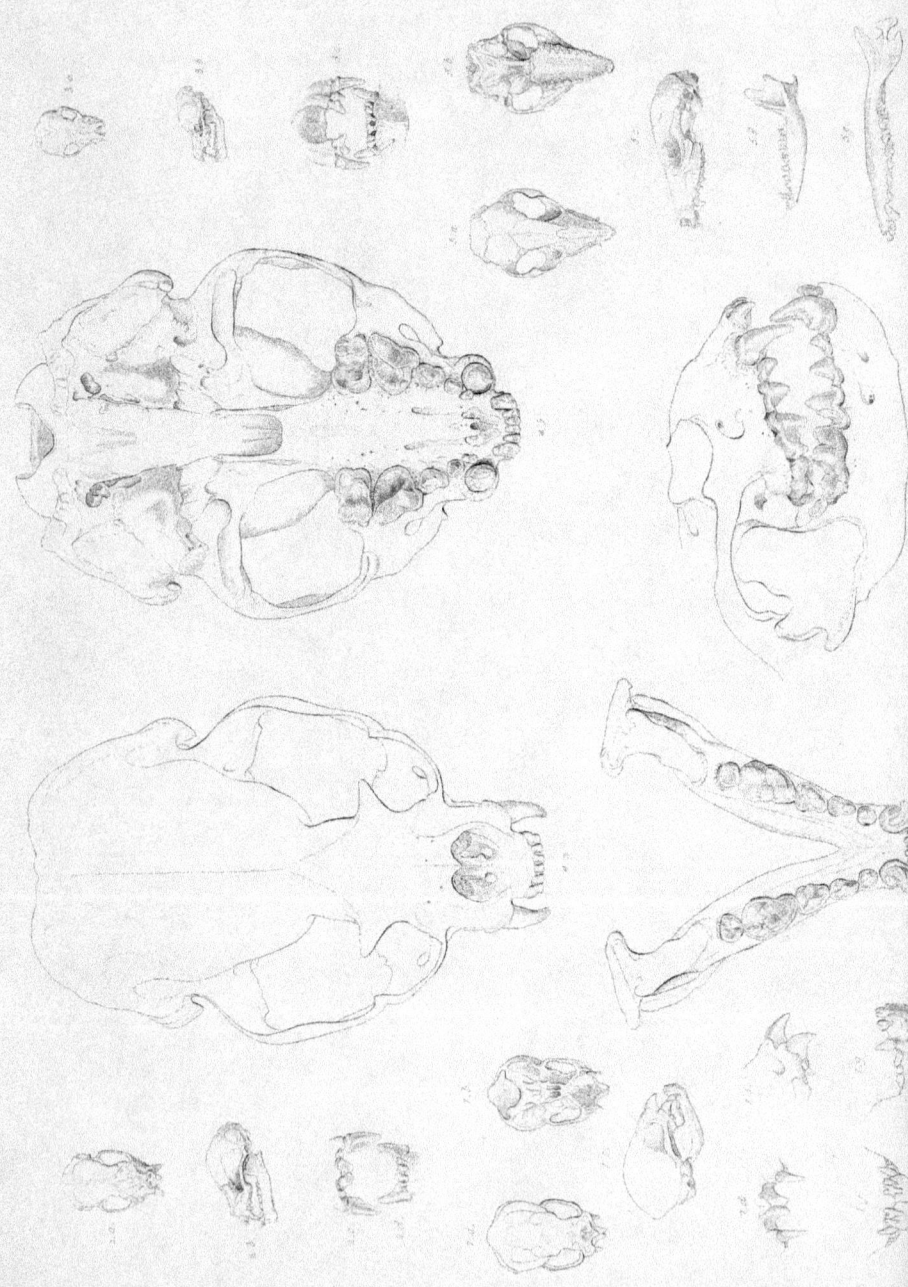

INDEX TO THE SPECIES.